Corrosion, Wear, Fatigue, and Reliability of Ceramics

Corrosion, Wear, Fatigue, and Reliability of Ceramics

A Collection of Papers Presented at the 32nd International Conference on Advanced Ceramics and Composites January 27–February 1, 2008 Daytona Beach, Florida

Editors

Jonathan Salem
Edwin R. Fuller

Volume Editors

Tatsuki Ohji
Andrew Wereszczak

The American Ceramic Society

WILEY

A John Wiley & Sons, Inc., Publication

Published by John Wiley & Sons, Inc., Hoboken, New Jersey.
Published simultaneously in Canada.

For general information on our other products and services or for technical support, please contact our Customer Care Department within the United States at (800) 762-2974, outside the United States at (317) 572-3993 or fax (317) 572-4002.

Wiley also publishes its books in a variety of electronic formats. Some content that appears in print may not be available in electronic format. For information about Wiley products, visit our web site at www.wiley.com.

Library of Congress Cataloging-in-Publication Data is available.

ISBN 978-0-470-34493-4

10 9 8 7 6 5 4 3 2 1

Contents

FATIGUE, WEAR, AND CREEP

RELIABILITY, NDE, AND FRACTOGRAPHY

Preface

This volume contains papers presented in the Mechanical Behavior and Structural Design of Monolithic and Composite Ceramics symposium of the 32nd International Conference & Exposition on Advanced Ceramics & Composites held on January 27-February 1, 2008 at Daytona Beach, Florida.

This long-standing symposium received presentations on a wide variety of topics providing the opportunity for researchers in different areas of related fields to interact. This volume emphasizes some practical aspects of real-world engineering applications of ceramics such as corrosion, fatigue, wear, reliability analysis, and fractography as associated with systems ranging from thermoelectric devices to solid oxide fuel cells.

The papers from the symposium represented research from 20 countries and demonstrate the worldwide interest in the properties and design of ceramic materials and components. The organization of the symposium and the publication of this proceeding were possible thanks to the professional staff of The American Ceramic Society and the tireless dedication of many Engineering Ceramics Division members. We would especially like to express our sincere thanks to the symposia organizers, session chairs, presenters and conference attendees, for their efforts and enthusiastic participation in the vibrant and cutting-edge symposium.

Jonathan Salem
NASA Glenn Research Center

Edwin R. Fuller
National Institute of Science and Technology

Introduction

Organized by the Engineering Ceramics Division (ECD) in conjunction with the Basic Science Division (BSD) of The American Ceramic Society (ACerS), the 32nd International Conference on Advanced Ceramics and Composites (ICACC) was held on January 27 to February 1, 2008, in Daytona Beach, Florida. 2008 was the second year that the meeting venue changed from Cocoa Beach, where ICACC was originated in January 1977 and was fostered to establish a meeting that is today the most preeminent international conference on advanced ceramics and composites

The 32nd ICACC hosted 1,247 attendees from 40 countries and 724 presentations on topics ranging from ceramic nanomaterials to structural reliability of ceramic components, demonstrating the linkage between materials science developments at the atomic level and macro level structural applications. The conference was organized into the following symposia and focused sessions:

Symposium 1	Mechanical Behavior and Structural Design of Monolithic and Composite Ceramics
Symposium 2	Advanced Ceramic Coatings for Structural, Environmental, and Functional Applications
Symposium 3	5th International Symposium on Solid Oxide Fuel Cells (SOFC): Materials, Science, and Technology
Symposium 4	Ceramic Armor
Symposium 5	Next Generation Bioceramics
Symposium 6	2nd International Symposium on Thermoelectric Materials for Power Conversion Applications
Symposium 7	2nd International Symposium on Nanostructured Materials and Nanotechnology: Development and Applications
Symposium 8	Advanced Processing & Manufacturing Technologies for Structural & Multifunctional Materials and Systems (APMT): An International Symposium in Honor of Prof. Yoshinari Miyamoto
Symposium 9	Porous Ceramics: Novel Developments and Applications

Symposium 10   Basic Science of Multifunctional Ceramics
Symposium 11   Science of Ceramic Interfaces: An International Symposium
                 Memorializing Dr. Rowland M. Cannon
Focused Session 1 Geopolymers
Focused Session 2 Materials for Solid State Lighting

Peer reviewed papers were divided into nine issues of the 2008 Ceramic Engineering & Science Proceedings (CESP); Volume 29, Issues 2-10, as outlined below:

- Mechanical Properties and Processing of Ceramic Binary, Ternary and Composite Systems, Vol. 29, Is 2 (includes papers from symposium 1)
- Behavior and Reliability of Ceramic Macro and Micro Scale Systems, Vol. 29, Is 3 (includes papers from symposium 1)
- Advanced Ceramic Coatings and Interfaces III, Vol. 29, Is 4 (includes papers from symposium 2)
- Advances in Solid Oxide Fuel Cells IV, Vol. 29, Is 5 (includes papers from symposium 3)
- Advances in Ceramic Armor IV, Vol. 29, Is 6 (includes papers from symposium 4)
- Advances in Bioceramics and Porous Ceramics, Vol. 29, Is 7 (includes papers from symposia 5 and 9)
- Nanostructured Materials and Nanotechnology II, Vol. 29, Is 8 (includes papers from symposium 7)
- Advanced Processing and Manufacturing Technologies for Structural and Multifunctional Materials II, Vol. 29, Is 9 (includes papers from symposium 8)
- Developments in Strategic Materials, Vol. 29, Is 10 (includes papers from symposia 6, 10, and 11, and focused sessions 1 and 2)

The organization of the Daytona Beach meeting and the publication of these proceedings were possible thanks to the professional staff of ACerS and the tireless dedication of many ECD and BSD members. We would especially like to express our sincere thanks to the symposia organizers, session chairs, presenters and conference attendees, for their efforts and enthusiastic participation in the vibrant and cutting-edge conference.

ACerS and the ECD invite you to attend the 33rd International Conference on Advanced Ceramics and Composites (http://www.ceramics.org/daytona2009) January 18–23, 2009 in Daytona Beach, Florida.

TATSUKI OHJI and ANDREW A. WERESZCZAK, Volume Editors
July 2008

Corrosion

CORROSION RESISTANCE OF CERAMICS IN VAPOROUS AND BOILING SULFURIC ACID

C.A. Lewinsohn, H. Anderson, and M. Wilson
Ceramatec Inc.
Salt Lake City, UT, 84119

T. Lillo
Idaho National Laboratory,
Idaho Falls, ID

A. Johnson
University of Nevada – Las Vegas
Las Vegas, NV

Large amounts of hydrogen can be produced using thermochemical processes, such as the Sulfur-Iodide (SI) process for thermo-chemical decomposition of water. The success of the SI processes is dependent on the corrosion properties of the materials of construction. Ceramic materials are required for high temperature decomposition reactors, since the creep and oxidation properties of super-alloy materials remain problematic due to the extreme temperatures (900C) and corrosive environments. In cooperation with the U.S. Department of Energy and the University of Nevada, Las Vegas (UNLV), ceramic based micro-channel decomposer concepts are being developed and tested by Ceramatec, Inc.. In order to assess the viability of ceramic materials, extended high temperature exposure tests have been made to characterize the degradation of the mechanical strength and estimate the recession rates due to corrosion. Corrosion has been investigated in vapour environments and in boiling, liquid-sulfuric acid. The results of these corrosion studies will be presented with additional analysis including surface and depth profiling using high resolution electron microscopy.

INTRODUCTION

The SI process for production of hydrogen from water involves continuously decomposing sulfuric acid (H_2SO_4) into oxygen, SO_2, and water and generating fresh sulfuric acid by removing the oxygen and reacting the SO_2 and water with iodine. The iodine removes some hydrogen from the water to form hydrogen iodide (HI) that can then be decomposed to form hydrogen, which is captured, and iodine, which is recycled. The remaining water reacts with the SO_2 to form fresh sulfuric acid and the cycle continues. Utilization of the sulfur-iodine thermochemical cycle to produce hydrogen involves the decomposition of sulfuric acid at elevated temperatures: 850-950°C [1]. This decomposition step creates a potentially corrosive environment, the effect of which on potential materials of construction is poorly characterized.

Some studies of the corrosion behavior of ceramic materials in high-temperature environments containing sulfuric acid were performed in the late 1970's and early 1980s after the SI process was proposed as a means to generate hydrogen for use as an energy carrier. Studies performed by Irwin and Ammon of the Westinghouse Electric Corporation, in 1981, found that silicon and materials containing significant amounts of silicon, such as silicon carbide and silicon nitride, have the greatest resistance to attack by boiling sulfuric acid.[2] In 1979, Fernanda Coen-Porisini of the Commission of the European Communities JRC Ispra

Establishment performed corrosion studies in sulfuric acid at 800°C and found that alumina, mullite, and zirconia, which were the few ceramics tested, were unchanged or had only coatings on the surface while all the metals tested displayed considerable to severe corrosion[3]. T. N. Tiegs of Oak Ridge National Laboratory also tested the corrosion resistance in simulated decomposed sulfuric acid of SiC, Sialon, MgO, $ZrO_2(MgO)$ and $ZrO_2(Y_2O_3)$ in 1981[4]. Tiegs identified silicon carbide as the best material at 1000 and 1225°C in the simulated sulfuric acid decomposition environment. Tiegs recommended further testing for the SiC materials at conditions more representative of an actual sulfuric acid decomposition environment, that is, at temperatures of 800 to 900°C and pressures up to 3 MPa. More recently, Ishiyama et al. report the percent weight change and corrosion rate of samples resulting from 100 hours of exposure to high-pressure boiling sulfuric acid. As seen from the results, SiC was the most corrosion resistant followed by Si-SiC and then by Si_3N_4. Also in Ishiyama's overall rating of the materials after 1000 hours of exposure the three above mentioned materials were listed as all being the least affected by the long exposure.

Based on the findings of these studies, silicon carbide and silicon nitride were selected as candidate heat exchanger/decomposer materials warranting further corrosion testing. This paper will describe the results from a series of corrosion tests to investigate the influence of the concentration of sulfuric acid and temperature on the corrosion of candidate ceramic materials for compact microchannel heat exchangers. Weight change, surface analysis, microscopy and mechanical testing were used to evaluate the effects of corrosion on the candidate materials. Alumina samples were included in the test matrix to verify the superior corrosion resistance of silicon-based materials and to compare their behavior with a material without silicon.

METHODS
1 – Experimental Setup
Samples were exposed to sulfuric acid at three conditions of temperature and pressure:

1. 850-950°C, atmospheric pressure (0.8 bar)
2. 375-400°C, 14 bar
3. 400°C, atmospheric pressure (0.8 bar)

The corrosive environments for these exposure testing were selected to mimic the decomposition and boiling environments of sulfuric acid in the SI process. In the SI process, decomposition will occur under pressure of up to 80 bar. Therefore, the second experimental setup was used to begin to investigate the effects of pressure on material behavior.

The high-temperature, sulfuric acid vapour corrosion test setup consists of a long quartz tube partially housed inside a split tube furnace. The long quartz tube itself holds three large quartz cups and three small quartz cups as displayed in Figure 1. Starting at the top is a large quartz cup filled with quartz chips which acts as an evaporator and gas preheater. Below the evaporator cup sit the three small cups that hold the samples. Below the three sample cups are two large condenser cups, the top of which is filled with Zirconia media and the bottom with SiC media. The long quartz tube is capped on top by a solid Teflon manifold with a pliable Teflon gasket so that the sulfuric acid vapor and decomposition products stay trapped in the tube. This manifold is fitted with gas (air/oxygen) feed and a liquid (sulfuric acid) feed. In addition, the condensate is collected and disposed of in an appropriate waste barrel.

ASTM Standard C 1161-02C bend bars were scribed, weighed and randomly positioned within three small sample cups for sulfuric acid exposure. The simulated conditions were 60% H_2SO_4, 30% H_2O and 10% air at 900°C. These sample cups were then loaded into the quartz tube and the furnace was heated up to 900°C with flowing argon gas. Once at temperature, the simulated sulfuric acid environment was attained by switching over to air or oxygen from argon and by dripping in the acid solution. At predetermined intervals (100, 200, 500 and 1000 hours), samples were removed, weighed and fractured according to ASTM Standard C 1161-02C procedures. Table I describes the materials that were tested.

Figure 1.   Sulfuric acid vapour exposure apparatus.

Table I

Material	Fabrication Process	Vendor	Identifier
Silicon Nitride	Hot Pressed	Ceradyne	SN-HP
Silicon Nitride	Gas Pressure Sintered	Ceradyne	SN-GP
Silicon Carbide	Pressureles s Sintered	Morgan	SiC-PS
Silicon Carbide	Tape Laminated	Ceramatec	SiC-LS

The intermediate temperature, high pressure corrosion apparatus was designed by Dr. T. Lillo at The Idaho National Laboratory. Various precautions were taken to minimize contact between the liquid sulfuric acid or acid vapours and the pressure containment vessel which is made of C-276, a metallic alloy. The experiment setup is shown in Figure 2. The typical volume of sulfuric acid in the experiment was 60 ml while the surface area of the samples was on the order of 160 mm^2, for a ratio of sulfuric acid volume to specimen surface area of 0.37 mL/mm^2 (ASTM G 31 indicates a minimum ratio of 0.2 mL/mm^2). The vessel was purged with UHP argon by pressurizing the vessel to approximately 500 psi and then releasing the pressure to approximately 50 psi. This was repeated three times to remove most of the air from the vessel prior to heating. At this time it is not known to what extent air/oxygen will be present in the system. At the target temperature of 375°C very little oxygen is expected to be present due to the large amount of activated carbon in our corrosion testing system. Heating to temperature took 1-2 hours and then was stable ($\pm$ 5°C of set point) throughout the experiment.

The sample was tested in the as-received condition, other than a brief ultrasonic agitation in ethanol to remove any residual grease and/or loose particles. The sample dimensions were measured to $\pm$0.01 mm and weighed to $\pm$0.00008 grams. The vessel was cooled after exposure experiments lasting up to 698 hours, opened and the sample rinsed, dried and weighed to yield a corrosion rate, according to ASTM G 31. The sample was then loaded back into the vessel with fresh acid and the experiment was continued. A total exposure time of 300-700 hours was targeted for each sample.

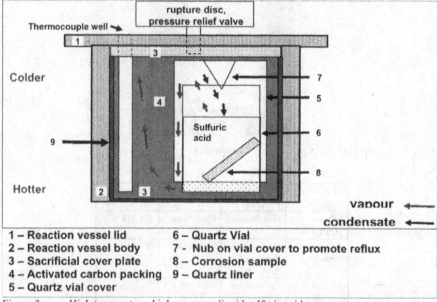

Figure 2.   High temperature, high pressure, liquid sulfuric acid exposure apparatus.

Experiments in boiling sulfuric acid, at atmospheric pressure, were conducted in a glass, reflux unit shown in Figure 3. The unit consisted of a heated flask and a reflux condenser cooled by recirculating, chilled water. ASTM Standard C 1161-02C bend bars were scribed, weighed and randomly positioned in a quartz fixture with three levels. Samples on the lowest level were submerged in sulfuric acid; the upper, only in contact with vapour. On the middle level, the interface between the liquid sulfuric acid and the vapour was at approximately the mid-plane of the samples. After 100 hours of exposure the samples were removed, wighed again, fractured, and examined by microscopy.

Figure 3.   Boiling sulfuric acid apparatus

RESULTS AND DISCUSSION
Vapour exposure

As reported earlier, all of the materials experienced an increase in weight as a result of exposure to a sulfuric acid decomposing environment at 900 °C. X-ray photoelectron spectroscopy (XPS) and Energy dispersive X-ray (EDX) analysis on a scanning electron microscope (SEM) indicated that the reaction products were silica. After exposure, both the silicon carbide and silicon nitride specimens exhibited a slight increase in strength, presumably due to the blunting of surface flaws by the formation of silica.

High-pressure, liquid exposure

As expected from previous corrosion experiments on other Si-based materials, the measured corrosion rates were relatively low. The weight loss behavior and calculated corrosion rate as a function of time for two silicon carbide specimens made using the materials and processes for making the proposed decomposition reactors are shown in Figure 4. A logarithmic curve fits the corrosion rate data best. The experimental data in Figure 4 suggests a very slightly decreasing corrosion rate with time. By normalising the weight change to the surface area of the specimens, recession rates of 125 μm/year and 62 μm/year were calculated for the silicon carbide and silicon nitride materials, respectively.

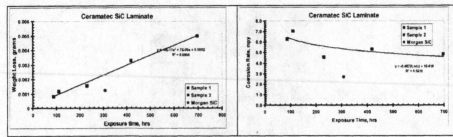

Figure 4.   Corrosion behavior of the Ceramatec SiC laminated samples. The solid line is fitted using the data from both samples.

Figure 5.   Comparison of exposed (375°C, 14 bar, 230 hours in 96 wt% H_2SO_4) and as-received Ceramatec Laminated SiC samples, left and right respectively.

Figure 6.   Specimen surface after 698 h of exposure (375°C, 14 bar, 230 hours in 96 wt% H_2SO_4).

The optical micrographs of the exposed samples provide evidence for simple dissolution. A comparison between an unexposed sample and the sample exposed for 230 hours is shown in

Fig. 5. The exposed sample retains many of the machining marks seen in the unexposed sample. However, a slight change in color and reflectivity indicate some level of reactivity, although the attack is not extensive. After 420 hours the machining marks have disappeared and after 698 hours there is little change in the macroscopic surface appearance. At higher magnification, however, needle-like grains are clearly visible, Fig. 6. There is no evidence of the formation of corrosion products or corrosion layer which would result in a decreasing corrosion rate with exposure time. Therefore, it appears this material degrades by simple dissolution and the corrosion rate can be expected to be nearly constant and independent of exposure time.

Boiling, liquid exposure
Silicon carbide specimens, made at Ceramatec, Inc., and gas-pressure sintered silicon nitride, manufactured by Ceradyne, Inc., were exposed for 100 h in the boiling, liquid exposure apparatus (Figure 3). The measured strength values and the calculated corrosion rates of the specimens are shown in Table II. The data indicate that corrosion is most severe in the boiling, liquid sulfuric acid. The corrosion rates of samples at the interface of the liquid and vapour, or completely in the vapour, exhibit much lower corrosion rates than those submerged in the liquid. It is possible that the formation of a silica film by oxidation in air inhibits corrosion by sulfuric acid. The silicon nitride specimens exposed to the sulfuric acid vapour, however, exhibited a degradation in strength due to the exposure, contrary to the results of exposure t sulfuric acid vapour at higher temperatures. Microscopy and surface analysis will be performed to understand this behavior.

Table II

Material, conditions	Avg. Strength	95% Confidence Interval	Corrosion Rate
Ceramatec SiC			
untreated	403 MPa	21 MPa	
boiling liquid, submerged	378 MPa	40 MPa	523 μm/yr
boiling liquid/vapour interface	396 MPa	44 MPa	1 μm/yr
vapour	381 MPa	62 MPa	2 μm/yr
Ceradyne Si_3N_4			
untreated	708 MPa	36 MPa	
boiling liquid, submerged	717 MPa	49 MPa	299 μm/yr
boiling liquid/vapour interface	707 MPa	27 MPa	70 μm/yr
vapour	325 MPa	60 MPa	59 μm/yr

SUMMARY
In summary, silicon-based ceramics offer potential for application in sulfuric acid decomposition environments, such as those required in the thermochemical generation of hydrogen using the Sulfur-Iodide process. In vaporous environments, at elevated temperatures, exposure to sulfuric acid containing environments caused silica films to form on the surfaces of the materials and subsequently an apparent increase in the mass and flexural strength of the materials. The highest rates of corrosion, leading to material loss, occurred in boiling liquid sulfuric acid, exposing material to liquid sulfuric acid at higher temperatures and pressure was less severe. It is likely that pressure retards the rate of corrosion by increasing the concentration of products and shifting equilibrium towards the reactants. The only condition leading to the degradation of strength was exposure of gas-pressure sintered silicon nitride to sulfuric acid

vapour above boiling sulfuric acid. Additional microscopy and surface analysis is required to understand the mechanisms of this behavior.

REFERENCES
[1] *Nuclear Hydrogen R&D Plan*. Department of Energy, Office of Nuclear Energy, Science and Technology. (2004, March). Retreived January 17, 2006, from http://www.hydrogen.energy.gov/pdfs/nuclear_energy_h2_plan.pdf
[2] Irwin, H.A., Ammon, R. L. *Status of Materials Evaluation for Sulfuric Acid Vaporization and Decomposition Applications*. Adv. Energy Syst. Div., Westinghouse Electric Corp., Pittsburg, PA, USA. Advances in Hydrogen Energy (1981), 2(Hydrogen Energy Prog., Vol. 4), 1977-99.
[3] Coen-Porisini, Fernanda. *Corrosion Tests on Possible Containment Materials for H2SO4 Decomposition.* Jt. Res. Counc., ERATOM, Ispra, Italy. Advances in Hydrogen Energy (1979), 1(Hydrogen Energy Syst., Vol. 4), 2091-112.

[4] Tiegs, T. N. (1981, July). *Materials Testing for Solar Thermal Chemical Process Heat*. Metals and Ceramics Division, Oak Ridge National Laboratory. Oak Ridge, Tennessee. ONRL/TM-7833, 1-59.

THERMOCOUPLE INTERACTIONS DURING TESTING OF MELT INFILTRATED CERAMIC MATRIX COMPOSITES

Ojard, G[2]., Morscher, G[3]., Gowayed, Y[4]., Santhosh, U[5]., Ahmad J[5]., Miller, R[2]. and John, R[1].

[1] Air Force Research Laboratory, AFRL/MLLMN, Wright-Patterson AFB, OH
[2] Pratt & Whitney, East Hartford, CT
[3] Ohio Aerospace Institute, Cleveland, OH
[4] Auburn University, Auburn, AL
[5] Research Applications, Inc., San Diego, CA

ABSTRACT:
As high performance ceramic matrix composite systems, such as Melt Infiltrated (MI) SiC/SiC, are being considered for advanced gas turbine engine applications, the characterization of the material becomes more important. A series of tests were conducted where Pt and Ni sheathed Pt thermocouples were used to monitor temperature for short and long duration fast fracture, fatigue and creep tests. While it is known that Si forms eutectics with Pt and Ni, this was initially not considered an issue. But since MI SiC/SiC composite achieves much of its performance from the infiltrated phase of Silicon (for high conductivity and low porosity), it was felt that further study of possible interactions of the Si phase has to be considered.

Post-test real-time X-ray inspection of the mechanical time-dependent and time-independent testing revealed that the extent of alloying into the sample was greater than anticipated and in some cases extended throughout the entire gage section of the tensile bars. It was concluded that the interactions were limited to the Si Phase of the material and that there was no difference between samples affected by alloying versus those that did not. Additionally, an experimental approach was taken to limit the extent of the thermocouple interaction. The results of this study and the approach will be presented and discussed.

INTRODUCTION
There is ever increasing interest in Ceramic Matrix Composites (CMCs) due to the fact that mechanical properties are relatively constant with temperatures up to the maximum use temperature [1]. This is shown in the interest of CMCs for extended high temperature use where superalloys are usually considered [2,3].
As characterization of this class of material proceeds to enable such applications, interactions present during the characterization become important. Most mechanical testing uses strain gauges, extensometry and or thermocouples. Strain gauges are normally used at room temperature where interactions can normally be considered to be absent and will not be discussed further. Most extensometers use alumina rods, which can be considered inert for the CMC class of material [4]. In addition, extensometers have point interactions with the specimen minimizing the area of contact. It should be noted that the extensometer could be influenced by the presence of porosity near the contact point during testing that would show up as discontinuities in strain [5] but this paper is focused on impact to the sample and not resulting testing impacts.
A main part of characterization is knowing the temperature at which the characterization is occurring. There are two ways to know the temperature: thermocouples (TC) and optical pyrometry. Both of these approaches were used by the authors in characterization and testing but the vast amount of the work has been with thermocouples. Due to the maximum temperature

range of interest (typically 1204°C), the thermocouples used consisted of Pt with or without Ni sheathing. In general CMC testing, there should be no issues with using thermocouples since the materials are usually inert: oxide, silicon carbide or silicon nitride matrix composite systems.

This could be an issue with the high performance CMC system generically referred to as the Melt Infiltrated CMC where a silicon based alloy is melted into the material as part of the matrix to increase through thickness thermal conductivity as well as decrease levels of porosity. Si is known to have eutectic interactions with Pt and Ni as well as other metals. (Pt and Ni are specifically pointed out since they are present in thermocouples at the highest percentages.) In this paper, a series of samples were exposed at elevated temperature for long periods of time under load for characterization purposes: creep and dwell fatigue. After testing, real time X-ray was undertaken on select samples to note if there was interaction or not. In addition, testing was reviewed to see if there was any affect on the test results.

PROCEDURE

Material Description

The material chosen for the study was the Melt Infiltrated SiC/SiC CMC system, which was initially developed under the Enabling Propulsion Materials Program (EPM) and is still under further refinement at NASA-Glenn Research Center (GRC). This material system has been systematically studied at various development periods and the most promising was the 01/01 Melt Infiltrated iBN SiC/SiC (01/01 is indicative of the month and year that development was frozen) [6]. There is a wide set of data from NASA for this system as well as a broad historic database from the material development [7]. This allowed a testing system to be put into place to look for key development properties which would be needed from a modeling effort and would hence leverage existing data generated by NASA-GRC.

The Sylramic® fiber was fabricated by DuPont as a 10 μm diameter stochiometric SiC fiber and bundled into tows of 800 fibers each. The sizing applied was polyvinyl alcohol (PVA). For this study, the four lots of fibers, which were used, were wound on 19 different spools. The tow spools were then woven into a 5 harness satin (HS) balanced weave at 20 ends per inch (EPI). An in-situ Boron Nitride (iBN) treatment was performed on the weave (at NASA-GRC), which created a fine layer of BN on every fiber. The fabric was then laid in graphite tooling to correspond to the final part design (flat plates for this experimental program). All the panels were manufactured from a symmetric cross ply laminate using a total of 8 plies. The graphite tooling has holes to allow the CVI deposition to occur. At this stage, another BN layer was applied. This BN coating was doped with Si to provide better environmental protection of the interface. This was followed by SiC vapor deposition around the tows. Typically, densification is done to about 30% open porosity. SiC particulates are then slurry cast into the material followed by melt infiltration of a Si alloy to arrive at a nearly full density material. The material at this time has less than 2% open porosity. Through this process, 15 panels were fabricated in 3 lots of material. Typical cross sections of this material are shown in Figure 1 showing the material phases.

After fabrication, all the panels were interrogated by pulse echo ultrasound (10 MHz) and film X-ray. There was no indication of any delamination and no gross porosity was noted in the panels. In addition, each panel had 2 tensile bars extracted for witness testing at room temperature. All samples tested failed above a 0.3% strain to failure requirement. Hence, all panels were accepted into the testing effort.

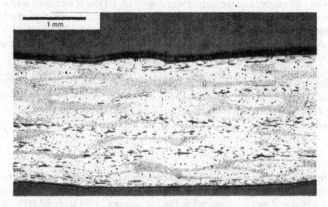

a) overall cross section

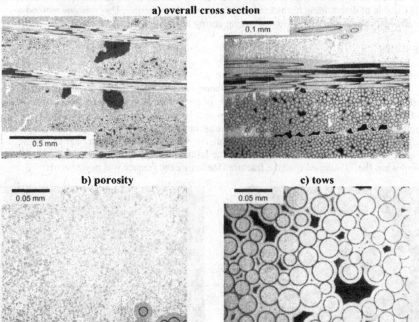

b) porosity   c) tows

d) SiC particulate with Si   e) Interface coating (BN)
Figure 1. MI SiC/SiC Microstructure Images

Thermocouples

Two types of thermocouples were used to monitor the temperature during testing: Type R and Type K. The vast majority of the testing was done with a Type R thermocouple while the long-term creep testing was done with a Type K thermocouple with the addition of a Ni sheath. A type R thermocouple is a bi-metallic joint between a Platinum −13% Rhodium alloy and Platinum [8]. A type K thermocouple is a bi-metallic joint between a Nickel-Chromium alloy and a Nickel-Aluminum alloy [8]. The thermocouples were held in place by wires.

Mechanical Testing - Durability Testing (Time-dependent)

The mechanical testing of MI SiC/SiC has been reported previously [9,10]. This work has been focused on the long-term durability response of the material under different loads and times: creep and dwell fatigue. Most of the testing was done at 1204°C with limited testing at 815°C.

Characterization - Real Time X-ray

As an aid in understanding the material, after creep and dwell fatigue testing, samples were interrogated using a real time x-ray machine. This effort was done trying to understand some low modulus material and was instrumental in finding porosity in the material [11]. In addition, it is capable of determining high-density inclusions in the material. This was one method to determine the presence of alloying occurring during testing.

RESULTS

Determination of initial reactions

During some of the initial testing efforts, thermocouple interactions were noted when Scanning Electron Microscopy (SEM) analysis of the failure face was performed. This is shown in Figure 2 for both the Type R and Type K (with a Ni sheath) thermocouples. With a combination of using Energy Dispersive Spectroscopy (EDS) and consulting binary phase diagrams, it was determined that these were eutectic spheres of Pt-Si (Figure 2a) and Ni-Si (Figure 2b). In addition, since these were at the failure face, it was thought to be a post failure event when the TC slipped onto the fracture face before the furnace was powered off.

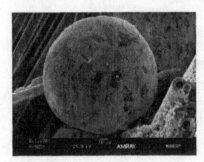 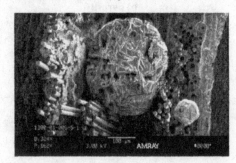

a) Type R TC used   b) Type K TC used (with a Ni sheath)
Figure 2. SEM Images of Failure Faces

Real Time X-ray Analysis

As noted previously, real time X-ray was used to document the presence of porosity in the sample gage region [11]. This was expanded to additional samples as an aid in determining sample outlier status [11]. During interrogation of samples, high-density inclusions were noted in the sample away from the failure face. The resultant real time X-ray images of the sample are shown in Figure 3. The dark inclusions in the material are regions that have been alloyed with Pt during the test. (The contrast is due to the greater absorption of the X-rays from the higher atomic number species present in the sample [12].) This sample was tested at 220.8 MPa for 1.3 hours at 1204°C suing a Type R TC. Energy Dispersive Spectroscopy (EDS) confirmed the presence of Pt in the SEM.

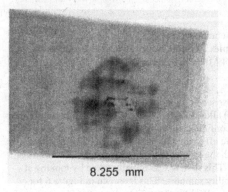

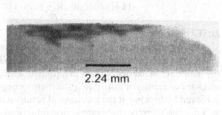

2.24 mm

8.255 mm

a) Normal view   b) Side View
Figure 3. Real Time X-ray of a Sample showing Interaction
(220.8 MPa creep test at 1204°C)

This was the first time that the presence of Pt (and in the case for the Ni sheathed Type K TC) was noted in the material and not at the failure face. Based on this finding, past testing was reviewed to note the full extent of the TC-Si interaction. The presence of alloying was greater than anticipated. This is shown in Figure 4 for two creep samples tested at 1204°C using a Type K TC with a Ni sheath. Figure 4 shows that the interaction zone is not time dependent and it can actually be greater in shorter duration tests.

a) Creep test at 110.4 MPa, 1204°C, 1269   a) Creep test at 165.6 MPa, 1204°C, 477 hours
hours – sample did not fail   – sample failed (Sample also shown in Fig. 2b)
Figure 4. Real Time X-ray of Two Samples Tested Under Different Stresses

For duplicate tests, the interaction was not always seen. For a series of 1 Hz fatigue tests (110.4 MPa and 1204°C) with a 2.286 mm through hole, the interaction was seen on one sample and not the other as shown in Figure 5. These samples were tested for 400,000 cycles and neither sample failed. The strain response of the material was identical.

a) No TC interaction   b) TC interaction
Figure 5. Real Time X-ray of Two Samples Tested Under Identical Conditions
(1 Hz Fatigue, R = 0.05, 110.4 MPa Peak Stress, 1204°C)

DISCUSSION
     As shown in the results, it is clear that the presence of a thermocouple (or the presence of a metal that can be alloyed with Si such as Pt or Ni) allows diffusion of another metal species into the material. What needs to be determined with confidence is this alloying affecting the mechanical results or not. (Or is the thermocouple affecting the testing outcome?) This was achieved by looking at the mechanical results with the insight gained by knowing if there was or was not any thermocouple interaction (alloying). This review was initially taken by looking at the creep response of some long-term creep durability samples. This is shown in Figure 6 for a series of samples that had a 4.572 mm through hole (W/d for the sample was set at 5) tested at 1204°C for 1,000 or 2,000 hours under a net section stress of 55.4 MPa.

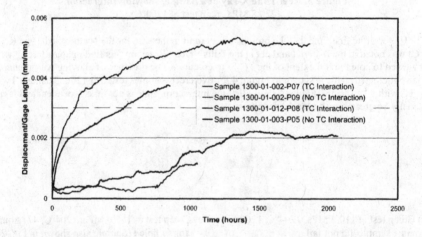

Figure 6. Long-term creep response of samples with and without thermocouple alloying

As can be seen in Figure 6, there was no differentiation between the displacement/gage length behavior with or without interaction. A review of other testing within the program for standard tensile samples and samples with holes showed similar response in that the presence of alloying did not differentiate the strain or displacement/gage length response. The residual strength testing done on samples that achieved run out backs up this conclusion. Two samples with a 2.286 mm through hole were tested at 110.4 MPa and 1204°C for 1,000 hours. One sample showed alloying to occur while the other did not. The residual tensile results plus a baseline sample are shown in Figure 7. As can be seen, the tensile curves show no differentiation.

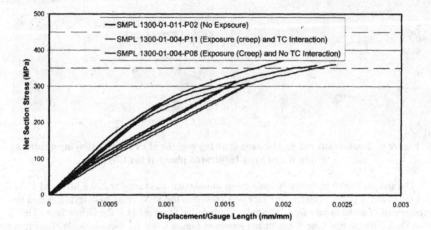

Figure 7. Residual tensile tests of samples with and without thermocouple alloying
(sample with through hole size of 2.286 mm, baseline curve shown for comparison purposes)

In addition, as further indication that there was no effect seen for the alloyed sample shown in Figure 7, posttest analysis showed that the hole dominated the failure location and not the region of alloying. This is shown in a real time X-ray image of the sample after the residual tensile test in Figure 8.

Figure 8. Real time X-ray image of sample with interaction showing that the alloying did not affect the failure location

Further analysis was done to note where the alloying was within the cross section of the material. A sample where a Pt thermocouple was used where alloying was known to occur from post real time X-ray analysis was sectioned. Back-scattered images were taken in the SEM and the presence of Pt was highlighted this way. It was seen that the Pt was staying in the melt-infiltrated phase of the material. This can be seen in Figure 9.

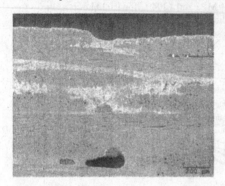 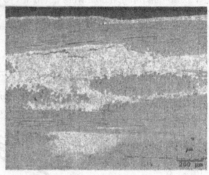

a)   b)

Figure 9. Back-scattered SEM image showing regions of Pt (white) alloying occurring within the Si Melt-Infiltrated phase of the CMC

The sample shown in Figure 9 was a creep sample that was tested at 220.8 MPa and 1204°C and failed at 1.3 hours. The fracture face was probed in the SEM using EDS trying to find the presence of Pt at the failure face. Multiple efforts failed to reveal Pt at the failure face. This is consistent with the real time X-ray image shown in Figure 8 that the thermocouple alloying is not affecting the testing results and or failure location. What is critical in Figure 9 and the back-scattered SEM work done is that it clearly shows that the alloying is not interacting with the CVI SiC phase of the material and additionally is not getting near the fiber interface coating. It would be expected that if the material were able to get to the interface that an effect would be seen.

It appears based on the work done to date that the thermocouple alloying is acting like a marker (tracer) within the system that is staying within the Si phase, which is around the CVI SiC coating of the fiber tows. This is clearly seen in Figure 10 where an edge on view of the sample taken in the real time X-ray machine shows that the alloying is present around the fiber tows. Since the temperature of testing where alloying occurred was always 1204°C and is above the eutectic point of 979°C (Si-Pt) and 993°C (Si-Ni), the alloying phase is highly mobile.

Figure 10. Side view of sample with alloying from real time X-ray
(Creep test at 110.5 MPa and 1204°C for 1269 hours)

As part of the effort to understand the interaction and to see if the effects can be minimized, a series of experiments were undertaken to note the presence of an interaction (or lack of an interaction) by having thermocouples placed on each side of the panel (from fabrication). As noted previously, the panel is fabricated in one of the last steps by placing Si on one side of the panel and then melting this through the panel. This results in one side of the panel being Si rich and the other side being Si poor. This is mainly seen in the appearance of the panel after fabrication. Thermocouples were placed on scrap material and then isothermally heat-treated at 1093°C for 12 hours. (While this was not the temperature used for testing, it was above the eutectic points as noted previously.) This was done for 3 samples of each condition: thermocouple on Si rich face and thermocouple on Si poor face. It was found that by placing the sample on the Si rich face that the interaction occurred on 2/3 of the samples while the thermocouples placed on the Si poor face had no interactions. The results of this experiment are shown in Figure 10 for the two cases studied.

a) thermocouple on Si rich face   b) thermocouple on Si poor face

Figure 10. Results of interrogating the two sides of the panel showing that the reaction can be seen when the thermocouple is placed on the Si rich face

CONCLUSION

It was seen that the thermocouple did interact with the Si phase of the Melt Infiltrated SiC/SiC composite during testing. Considering that the standard thermocouple materials form eutectics with Si, this is no longer surprising. Reviewing the data generated with insight from X-ray interrogation of the samples, it was seen that the presence of the alloying within regions of the material were not affecting the results. Additionally, there was no effect seen on the failure location of the material. This is further confirmed by the investigative SEM work done where EDS and back-scattered images show that the alloying is staying strictly within the Si phase of the CMC and is not penetrating the CVI SiC around the fiber tows and not attacking the fiber interface coating.

After the interaction was noted, the experimental work done by placing the thermocouples on the two different sides of the panels shows that the interaction can be reduced. By placing the thermocouple on the Si poor side of the panel, the presence of the interaction can be reduced. This should eliminate this as a concern but even with this, care should be taken to look at samples to assure that alloying still has not occurred since the limited testing here could not cover all the conditions that could exist on the Si poor face of the material (there could be local spots on the Si poor face that due to the panel fabrication could be rich in Si).

All of this work to date is for isothermal testing above the eutectic point. Micro-structural examination upon sample cooling did not show any eutectic phases formed considering the extent of the alloying seen in the real time X-ray. This may be due to the fact that sample cooling after testing was sufficiently fast such that the kinetics of formation of the phases could not occur. It is unknown at this time what would occur if the temperature were held below the eutectic for a sufficient time to allow the phases of the eutectic to form. Thermal cycling with alloying occurring that goes through the eutectic point may be a problem if the phases formed have different volumes. This is an area for concern in use and testing that should be studied. In the end, researchers should be aware of possible interactions when investigating this class of material (MI SiC/SiC) and take care in evaluating their results regardless of how the test is performed (such as the isothermal work presented here).

ACKNOWLEDGMENTS

The Materials & Manufacturing Directorate, Air Force Research Laboratory under contract F33615-01-C-5234 and contract F33615-03-D-2354-D04 sponsored portions of this work

REFERENCES
[1]Wedell, James K. and Ahluwalia, K.S., "Development of CVI SiC/SiC CFCCs for Industrial Applications" 39th International SAMPE Symposium April 11- 14, 1994, Anaheim California Volume 2 pg. 2326.
[2]Brewer, D., Ojard, G. and Gibler, M., "Ceramic Matrix Composite Combustor Liner Rig Test", ASME Turbo Expo 2000, Munich, Germany, May 8-11, 2000, ASME Paper 2000-GT-0670.
[3]Calomino, A., and Verrilli, M., "Ceramic Matrix Composite Vane Sub-element Fabrication", ASME Turbo Expo 2004, Vienna, Austria, June 14-17, 2004, ASME Paper 2004-53974.
[4]Harper, C.A., 2001, *Handbook of Ceramics, Glasses, and Diamonds*, McGraw-Hill, New York, Chap. 1.
[5]Ojard, Greg C. unpublished research.
[6]Calomino, A., NASA-Glenn Research Center, personal communication.
[7]J.A. DiCarlo, H-M. Yun, G.N. Morscher, and R.T. Bhatt, "SiC/SiC Composites for 1200°C and Above" *Handbook of Ceramic Composites,* Chapter 4; pp. 77-98 (Kluwer Academic; NY, NY: 2005)
[8] http://www.omega.com/Temperature/pdf/SPPL.pdf
[9]Ojard, G., Gowayed, Y., Chen, J., Santhosh, U., Ahmad J., Miller, R., and John, R., "Time-Dependent Response of MI SiC/SiC Composites Part 1: Standard Samples" to be published in Ceramic Engineering and Science Proceedings, 2007.
[10]Y. Gowayed, G. Ojard, J. Chen, R. Miller, U. Santhosh, J. Ahmad and John, R., "Time-Dependent Response of MI SiC/SiC Composites Part 2: Samples with Holes", to be published in Ceramic Engineering and Science Proceedings, 2007.
[11] Ojard, G., Miller, R., Gowayed, Y., Santhosh, U., Ahmad, J., Morscher, G., and John, R. "MI SiC/SiC Part 1 – Mechanical Behavior", 31st Annual Conference on Composites, Materials and Structures, Daytona Beach, FL.
[12] Halmshaw, R. 1987, *Non-destructive Testing*, Edward Arnold Ltd. London, Chap. 2.

OXIDATION RESISTANCE OF PRESSURELESS-SINTERED SiC-AlN-Re$_2$O$_3$ COMPOSITES OBTAINED WITHOUT POWDER BED

G.Magnani
ENEA
Dept. of Physics Tech. and New Materials
Bologna Research Center
Via dei Colli 16
40136 Bologna
Italy

F.Antolini, L.Beaulardi, F.Burgio, C.Mingazzini
ENEA
Dept. of Physics Tech. and New Materials
Faenza Research Center
Via Ravegnana 186
48018 Faenza (Ra)
Italy

ABSTRACT

Oxidation resistance of SiC-AlN-Re$_2$O$_3$ composites (SiC 50%wt-AlN 50%wt) pressureless-sintered with an innovative and cost-effective method is reported. Yttria, Lutetia, Ytterbia and Erbia were tested as sintering-aids in this pressureless-sintered process which does not require the use of the powder bed to protect samples during the heat treatments. Sintered density was always >95%T.D. and microstructure was mainly composed by 2H SiC-AlN solid solution with well-dispersed grain boundary phases. The sintered samples, after oxidation in the temperature range 1200-1500°C over a period of time of 200h, showed a parabolic oxidation behaviour with diffusion of oxygen through the silica-based surface and interfacial reactions silica-alumina and silica-rare-earth oxides as rate-determining steps.

INTRODUCTION

Silicon carbide (SiC) and aluminium nitride (AlN) form a 2H solid solution which has received considerable attention owing to its high potential for application in chemically aggressive environments[1-2]. The SiC-AlN composites were preferably sintered by hot pressing in inert[3] or nitrogen atmosphere[4] or in vacuum[5]. Gas pressure sintering (GPS) was also developed[6], while pressureless sintering with liquid-phase forming additives (LPS) was partially inhibited due to the large amount of evaporation loss associated with various chemical reactions[7]. In this case a powder bed was normally used to limit the effects of the decomposition of species like yttria, added as sintering aids to the SiC-AlN mixture, and alumina[8,9]. In a previous paper, we have already demonstrated that high density SiC-AlN ceramics can successfully be obtained by liquid-phase pressureless sintering without using a powder bed[10]. In this paper, the same process was applied to SiC-AlN composites containing different rare-earth oxides (Yb$_2$O$_3$, Er$_2$O$_3$, Lu$_2$O$_3$) as-sintering aids.

In addition, the evaluation of the application of ceramics to systems operating at high temperature depends on the mechanical reliability of the materials during thermal exposure to oxidizing conditions at high temperature. Therefore, it is required for this class of materials to learn oxidation behaviour at elevated temperature. On the basis of these considerations, the long term (200h) oxidation behaviour of SiC-AlN-Re$_2$O$_3$ ceramics was also analysed and compared to the high temperature resistance of analogous materials obtained with other sintering routes.

EXPERIMENTAL PROCEDURE

Commercially available α-SiC (UF10, H.C.Starck, Germany), AlN (Grade F, Tokuyama Soda, Japan), Y$_2$O$_3$ (H.C.Starck, Germany), Yb$_2$O$_3$, Er$_2$O$_3$ and Lu$_2$O$_3$ (Cerac Inc., USA) were used as starting powders. Characteristics of these powders are reported in Table I.

Batch of powder composed by 48%wt SiC, 48%wt AlN and 4%wt Re$_2$O$_3$ was wet-mixed in ethanol for 12 h using SiC grinding balls. After drying and sieving, the powder was compacted by die pressing at 67 MPa and subsequently was pressed at 150 MPa by CIP.

21

Sintering was performed in a graphite elements furnace in flowing nitrogen at 1 atm with green bodies put inside a graphite crucible without powder bed. Sintering was performed at 1950°C, while an annealing step was conducted at 2050°C. Thermal cycle was characterised by heating and cooling rate of 20-30°C/min and by dwell time of 0.5 h at the sintering temperature.

The bulk densities of the sintered samples were determined using the Archimede method. The microstructures were characterized using scanning electron microscopy (SEM), while X-ray diffractometry (XRD) using Cu$K\alpha$ radiation was performed in order to identify crystalline phases contained in the sintered and oxidized samples.

Oxidation experiments were carried out at different temperatures in the range 1200-1500°C over a period of 200 h in air. Rectangular pellets (18mm x 3mm x 3mm) were prepared from the bulk specimen with a diamond saw. After grounding to reduce superficial roughness, the specimens were cleaned in an ultrasonic bath and degreased with acetone and ethanol. Dried samples were then weighed and the exact dimensions were measured in order to calculate the surface area. The experiments were conducted in a furnace having molybdenum disilicide heating elements.

Table I. Characteristics of the starting powders

Powder	Purity (wt%)	Specific Surface Area (m²/g)	Particle size (μm)
α-SiC	>97.0	15.6	0.48
AlN	>98.0	3.3	0.1-0.5
Y₂O₃	99.9		<5
Lu₂O₃	99.9		<5
Er₂O₃	99.9		<5
Yb₂O₃	99.9		<5

RESULTS AND DISCUSSION

Density and microstructure

Bulk density, weight loss and crystalline phases of the sintered samples are reported in Table II. Residual porosity (<5%) was due to the weight loss associated to the formation of gaseous products of redox reactions between SiC and SiO₂, Al₂O₃ or Re₂O₃:

$$SiC + 2SiO_2 \Leftrightarrow CO_{(g)} + 3SiO_{(g)} \qquad (1)$$

$$SiC + Al_2O_3 \Leftrightarrow CO_{(g)} + SiO_{(g)} + Al_2O_{(g)} \qquad (2)$$

$$SiC + 2Re_2O_3 \Leftrightarrow CO_{(g)} + SiO_{(g)} + 4ReO_{(g)} \qquad (3)$$

Previous work focused on the study of LPS-SiC(Y₂O₃) clearly showed that the main contribution to the weight loss comes from reactions (1) and (2), while weight loss is not greatly influenced by the decomposition of yttria (reaction (3))[11]. Rare-earth oxides are the species most thermodynamically stable among those contained in the samples and unlikely to decompose under carbothermally reducing conditions at temperatures up to 2000°C[12]. This is not true in the case of SiC-AlN-Yb₂O₃ system where weight loss is greater than the other systems and bulk density is consequently lower. Same behaviour was reported by **Izhevskyi et al.**[13] **in liquid-phase sintered SiC with AlN and Yb₂O₃ as sintering-aids.**

On the other hand, rare-earth oxides participate to the formation of the grain boundary phases reported in Table II. These crystalline phases act as liquid phase that assists densification during pressureless sintering[10] and precipitates at the grain boundary during cooling.

Furthermore, SEM micrographs, obtained with back-scattered electrons and reported in Figure 1, put in evidence the bright spots of the liquid phase having a higher average atomic number and revealed that the grain boundary phases are homogenously distributed in the samples. This fact confirms that weight loss can be controlled during sintering even though a powder bed to protect the green bodies is not used[11].

Table II. Characteristics of the sintered samples (mp=main phase, sp= secondary phase, gbp=grain boundary phase, tr=trace)

Material	Bulk density (%T.D.)	Sintering weigth loss (%wt)	Crystalline phase
SiC-AlN-Y₂O₃	97	3.5	2H SiC-AlN (mp), α-SiC (sp), $Y_{10}Al_2Si_3O_{18}N_4$ (gbp)
SiC-AlN-Lu₂O₃	97	4.6	2H SiC-AlN (mp), α-SiC (sp), Lu_2O_3 (gbp), $Lu_2Si_2O_7$ (gbp), Cristobalite (tr)
SiC-AlN-Er₂O₃	97	3.4	2H SiC-AlN, α-SiC, Er_2O_3 (gbp), $Er_2Si_2O_7$ (gbp)
SiC-AlN-Yb₂O₃	95	9.1	2H SiC-AlN, α-SiC, Yb_2O_3 (gbp), $Yb_2Si_2O_7$ (gbp), Cristobalite (tr)

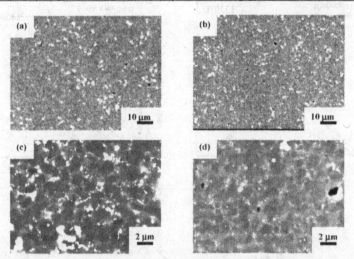

Figure 1: Back-scattered electrons images of pressureless-sintered a) SiC-AlN-Y₂O₃, b) SiC-AlN-Lu₂O₃, c) SiC-AlN-Er₂O₃-and d) SiC-AlN-Yb₂O₃ composites.

Oxidation behaviour

Figure 2 shows the relation between the square of the weight gain and the oxidation time for specimens oxidized in air at temperature between 1200°C and 1500°C. For all temperatures, oxidation kinetics were of the parabolic type. Thus, the oxidation behaviour is governed by the parabolic rate equation:

$$\Delta W^2 = kt \qquad (4)$$

where ΔW is the specific weight gain, k is the kinetic constant of parabolic oxidation and t is the oxidation time.

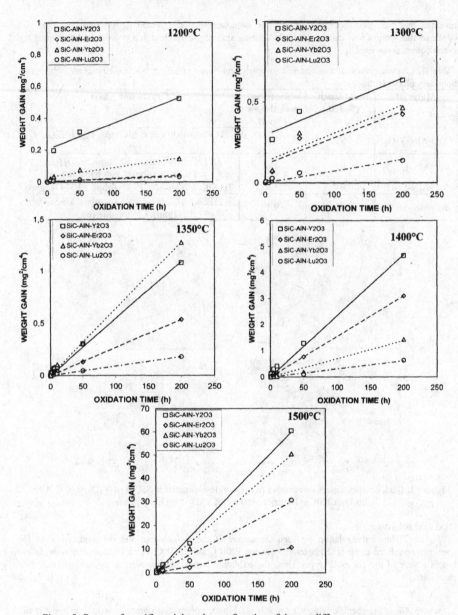

Figure 2: Square of specific weight gain as a function of time at different temperature.

The oxidized samples at 1200°C and 1300°C with Re=Y, Yb, Er appear to show different slopes as function of the oxidation time. This may be caused by the rapid increase of the thickness of rare-earth cations depletion zone beneath the oxidized layer with increasing oxidation time, which requires longer time for diffusion of additive cations, resulting in the decrease in oxidation rate with increasing time.

On the basis of Eq.4, parabolic rate constants for each temperature were determined from the slope of the straight lines reported in Figure 2 (parabolic rate constants for oxidation at 1200°C and 1300°C were determined in the range 10-200h).

The values of specific weight gains and parabolic rate constants for each SiC-AlN-Re₂O₃ composite are summarized in Table III together to the experimental data of other SiC-AlN composites manufactured in different conditions[14-20].

SiC-AlN-Y₂O₃ and SiC-AlN-Er₂O₃ respectively show the lowest and the highest oxidation resistance at 1500°C, whereas the comparison with other SiC-AlN materials is quite difficult due to the difference in the starting SiC-AlN weight ratio, in the temperature range and in the duration of oxidation treatments. Nevertheless, if experimental data of the materials sintered with the same rare-earth oxide are compared, our materials showed the best oxidation resistance together to the materials studied by Choi et al.[16] and Biswas et al.[20]. In addition, it is interesting to note that our materials presented specific weight gain (and parabolic rate constant) lower than hot isostatic pressed materials manufactured without sintering-aid[14] and SiC-AlN-Y₂O₃ composites pressureless-sintered with powder bed[15].

Parabolic oxidation behaviour of ceramics normally indicates that the rate-determining step is a diffusional process associated with the migration of ions[20]. In the case of solid state sintered SiC, Luthra[21] reviewed many studies on oxidation of SiC and concluded that, although most observed parabolic oxidation rate, mixed diffusion/reaction rate mechanism must be controlling the oxidation of SiC. On the basis of parabolic rate constants reported in Table III, we were able to determine the activation energy between 1200°C and 1500°C on the basis of the Arrhenius law:

$$k = k_0 \, exp\left(-\frac{Q}{RT}\right)$$ (5)

where k_0 is a material constant, Q is the oxidation activation energy, T is the absolute temperature and R is the gas constant. Recently, Biswas et al.[20] found high activation energies (250-560 KJ/mol) for the oxidation process in lutetia-doped SiC-AlN composites (Table III) and consequently suggested that oxidation proceeds not only by the diffusion of oxygen through the silica layer, but also by interfacial reactions between the growing silica layer and lutetia. The same conclusion was postulated by Guo et al.[18] who found a Q value of 600 KJ/mol for oxidation of liquid-phase sintered SiC with AlN and Er₂O₃ as sintering-aids (Table III).

The apparent oxidation activation energy, calculated from the slope of the straight line reported in Figure 3, was found to be in the range 386-500 KJ/mol (Table III), higher than the activation energy associated to the diffusion of oxygen both through the silica (150 KJ/mol)[20] and mullite layer (260-290 KJ/mol)[22].

These values suggest that not only the inward diffusion of oxygen through the silica-based surface oxide scale have to be considered as rate-controlling mechanism, but also the interfacial reaction between SiO₂ and Al₂O₃, oxidation products of SiC and AlN, which led to the formation of mullite:

$$SiC_{(s)} + \tfrac{3}{2} O_{2(g)} \Leftrightarrow SiO_{2(s)} + CO_{(g)}$$ (6)

$$4AlN_{(s)} + 3O_{2(g)} \Leftrightarrow 2Al_2O_{3(s)} + 2N_{2(g)}$$ (7)

$$2SiO_{2(s)} + 3Al_2O_{3(s)} \Leftrightarrow 3Al_2O_3.2SiO_{2(s)}$$ (8)

Table III. Comparison of the weight gain and parabolic rate constant (k) of the SiC-AlN composites developed in this study with those obtained by other authors.

Reference	Sintering	Additive	SiC/AlN weight ratio	Weight gain (mg/cm^2)	Parabolic rate constant (Kg^{2}m^{-4}s^{-1})	Q (KJ/mol)
This study	LPPS without powder bed	Y$_2$O$_3$	50/50	0.72 (1200°C.200h) 0.80 (1300°C.200h) 1.04 (1350°C.200h) 2.16 (1400°C,200h) 7.77 (1500°C,200h)	4.44x10^{-11}(1200°C) 5.00x10^{-11}(1300°C) 1.53x10^{-10}(1350°C) 6.53x10^{-10}(1400°C) 8.29x10^{-9}(1500°C)	386±98
		Er$_2$O$_3$	50/50	0.20 (1200°C.200h) 0.65 (1300°C.200h) 0.73 (1350°C.200h) 1.70 (1400°C.200h) 3.25 (1500°C.200h)	5.56x10^{-12}(1200°C) 4.44x10^{-11}(1300°C) 7.50x10^{-11}(1350°C) 4.03x10^{-10}(1400°C) 1.45x10^{-9}(1500°C)	410±31
		Yb$_2$O$_3$	50/50	0.38(1200°C.200h) 0.68 (1300°C.200h) 1.13 (1350°C.200h) 1.06(1400°C.200h) 7.09(1500°C.200h)	1.67x10^{-11}(1200°C) 5.00x10^{-11}(1300°C) 1.78x10^{-10}(1350°C) 1.53x10^{-10}(1400°C) 6.90x10^{-9}(1500°C)	408±95
		Lu$_2$O$_3$	50/50	0.18(1200°C.200h) 0.37 (1300°C.200h) 0.42 (1350°C.200h) 0.81 (1400°C.200h) 5.51 (1500°C.200h)	2.78x10^{-12}(1200°C) 1.94x10^{-11}(1300°C) 2.50x10^{-11}(1350°C) 8.89x10^{-11}(1400°C) 4.14x10^{-9}(1500°C)	500±96
Lavrenko et al.[14]	Hot pressing		50/50	1.5 (1350°,6h)	0.9x10^{-8} (1350°C)	
Sciti et al.[15]	LPPS with powder bed	Y$_2$O$_3$	22/78	2.79 (1350°C,30h) 10.75 (1400°C,30h)	7.3x10^{-9}(1350°C) 1.1x10^{-7}(1400°C)	n.d. n.d.
Choi et al.[16]	Hot pressing	Y$_2$O$_3$	97/3	1.99 (1400°C,192h)	5.28x10^{-10}(1400°C)	n.d.
		Er$_2$O$_3$	97/3	1.70 (1400°C,192h)	3.42x10^{-10}(1400°C)	n.d.
		Yb$_2$O$_3$	97/3	0.47 (1400°C,192h)	3.06x10^{-11}(1400°C)	n.d.
Pan et al.[17]	Hot pressing	Y$_2$O$_3$	85/15	3.00 (1370°C,30h)	n.d.	
Guo et al.[18]	Gas pressure sintering	-	97/3	0.247 (1200°C,200h) 0.704 (1300°C,200h) 4.426 (1400°C,200h)	n.d. n.d. n.d.	600
Biswas et al[19]	Gas pressure sintering	Lu$_2$O$_3$	95/5	0.18 (1200°C,100h) 0.39 (1400°C,100h) 2.25 (1500°C,100h)	1.0x10^{-11} 1.0x10^{-10} 4.5x10^{-9}	500±60
Biswas et al[20]	Gas pressure sintering	Lu$_2$O$_3$	96/4	0.16 (1200°C,100h) 0.53 (1400°C,100h) 1.52 (1500°C,100h)	5.43x10^{-12} 7.94x10^{-11} 6.44x10^{-10}	561±67
			97.5/2.5	0.20 (1200°C,100h) 0.57 (1400°C,100h) 1.24 (1500°C.100h)	8.72x10^{-12} 4.11x10^{-11} 4.24x10^{-10}	460±60
			99/1	0.35 (1200°C,100h) 0.40 (1400°C,100h) 1.13 (1500°C,100h)	3.33x10^{-11} 8.58x10^{-11} 3.59x10^{-10}	251±184

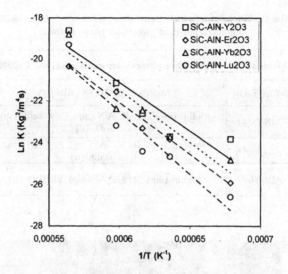

Figure 3: Arrhenius plot of parabolic rate constants for oxidation.

This mechanism was confirmed by means of XRD analysis performed on the oxidized surface after oxidation at 1500°C for 200h (Table IV). Mullite was always found as main oxidation product, without regard to the rare-earth oxides used. Silica could also react with Re_2O_3 to form rare-earth silicates ($Re_2Si_2O_7$):

$$2SiO_{2(s)} + Re_2 O_{3(s)} \Leftrightarrow Re_2 Si_2 O_{7(s)} \qquad (9)$$

In fact, rare-earth disilicates were also detected by XRD analysis (Table IV), while rare-earth oxides found in the as-sintered samples were no longer revealed after oxidation. $Re_2Si_2O_7$ forms typical acicular grains as reported in Figure 4. Therefore, cristobalite is simultaneously formed by reaction (6) and consumed by reactions (8) and (9). On the other hand, reaction (9) requires outward cation diffusion from the intergranular region. Consequently, an oxidation/reoxidation experiment was performed[23] in order to distinguish the rate controlling mechanism between the interfacial reaction to form mullite (inward diffusion of oxygen) and outward cation diffusion. The mass-gain curve for reoxidation substantially repeated the initial stage parabola confirming that the inward diffusion of oxygen through the protective surface oxide scale represents the rate control mechanism.

Table IV. Crystalline phases detected after oxidation at 1500°C for 200h
(mp=main phase, sp= secondary phase, tr=trace)

Material	Crystalline phases after oxidation (1500°Cx200h)
SiC-AlN-Y$_2$O$_3$	Mullite (mp), 2H SiC-AlN (sp)
SiC-AlN-Lu$_2$O$_3$	Mullite (mp), 2H SiC-AlN (mp), Cristobalite (sp), Lu$_2$Si$_2$O$_7$ (tr)
SiC-AlN-Er$_2$O$_3$	Mullite (mp), 2H SiC-AlN (sp), Er$_2$Si$_2$O$_7$ (sp), Cristobalite (tr)
SiC-AlN-Yb$_2$O$_3$	Mullite (mp), 2H SiC-AlN (sp), Yb$_2$Si$_2$O$_7$ (tr),

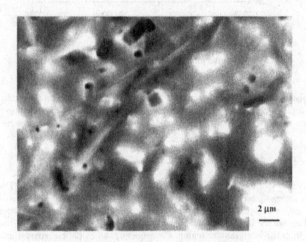

Figure 4: SEM image of the oxidized surface (1500°C, 200h) of SiC-AlN-Yb$_2$O$_3$ showing the acicular grains of Yb$_2$Si$_2$O$_7$.

Then, the cross-sections of the oxidized samples were examined by SEM in order to evaluate thickness and porosity of the protective oxide scale (Figure 5). SiC-AlN-Y$_2$O$_3$ ceramic shows an oxide layer of 180-200 µm unable to act as protective barrier against oxygen diffusion (see EDS oxygen X-ray map reported in Figure 4). On the contrary, SiC-AlN-Er$_2$O$_3$ materials present a thinner (80 µm) and more compact scale which is able to improve the oxidation resistance of the pressureless sintered 50%wtSiC-50%wtAlN composites.

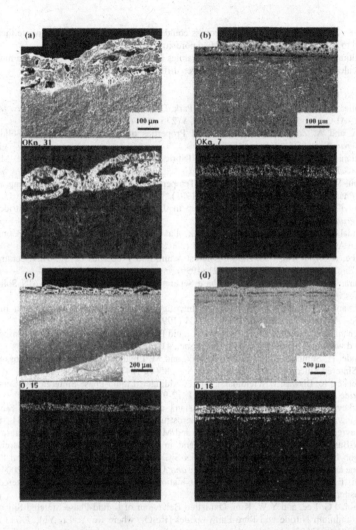

Figure 5: SEM micrographs and EDS oxygen X-ray maps of pressureless-sintered a) SiC-AlN-Y$_2$O$_3$, b) SiC-AlN-Lu$_2$O$_3$. c) SiC-AlN-Er$_2$O$_3$-and d) SiC-AlN-Yb$_2$O$_3$ composites after oxidation at 1500°C for 200h.

CONCLUSION

SiC-AlN-Re$_2$O$_3$ (Re=Y, Yb, Er, Lu) composites could be successfully pressureless-sintered without protective powder bed and with different rare-earth oxides as sintering aid. Microstructure is composed by fine grains of 2H SiC-AlN solid solution with liquid phase located at the grain boundary.

Weight loss associated to the sintering process could be controlled with an appropriate thermal cycle which led to a sintered body with low residual porosity.

Oxidation behaviour of SiC-AlN-Re$_2$O$_3$ ceramics is controlled by the formation of a mullite-based surface scale that protects samples against oxygen diffusion inwards.

REFERENCES

[1] I.B. Cutler, P.D. Miller, W. Rafaniello, H.K. Park, D.P. Thompson, and K.H. Jack, New Materials in the Si-C-Al-O-N and Related System, *Nature*, V(275), 434 (1978).

[2] R. Ruh, and A. Zangvil, Composition and Properties of Hot-Pressed SiC-AlN Solid Solution, *J.Am.Ceram.Soc.*, **65**(5), 260-65 (1982).

[3] W. Rafaniello, K. Cho, and V. Virkar,, Fabrication and Characteristics of SiC-AlN Alloys, *J.Mater.Sci.*, , **16**, 3479-88 (1981).

[4] Z.C. Jou, V. Virkar, and A.R Cutler, High Temperature Creep of SiC Densified using a Transient Liquid Phase, *J. Mater. Res.*, **6**(9), 1945-49 (1991).

[5] A. Zangvil and R. Ruh, "Phase Relationship in the Silicon Carbide-Aluminum Nitride System", *J.Am.Ceram.Soc.*, **71**(10), 884-90 (1988).

[6] S. Mandal, K.K. Dhargupta., and S. Ghatak, Gas Pressure Sintering of SiC-AlN Composites in Nitrogen Atmosphere, *Ceram.Int.*, **28**, 145-51 (2002).

[7] R.R. Lee, and W. Wei, Fabrication, Microstructure and Properties of SiC-AlN Ceramic Alloys, *Ceram.Eng.Sci.Proc.*, **11**(7-8), 1094-1121 (1990).

[8] M. Miura, T. Yogo, and S.I. Hirano, Phase Separation and Toughening of SiC-AlN Solid-Solution Ceramics, *J.Mater.Sci.*, **28**, 3859-65 (1993).

[9] J.F. Li, and R. Watanabe, Pressureless Sintering and High-Temperature Strength of SiC-AlN Ceramics, *J. Ceram. Soc.Japan*, **102**(8), 727-31 (1994).

[10] G. Magnani, and L. Beaulardi, Properties of Liquid Phase Pressureless Sintered SiC–Based Materials Obtained without Powder Bed, *J.Aus.Ceram.Soc.*, **41**(1), 31-6 (2005).

[11] T. Grande, H. Sommerset, E. Hagen, K. Wiik, and M. Einarsud, Effect of Weight Loss on Liquid-Phase-Sintered Silicon Carbide, *J.Am.Ceram.Soc.*, **80**(4), 1047-52 (1997).

[12] G. Rixecker, K. Biswas, I. Wieldmann, and F. Aldinger, Liquid-Phase Sintered SiC Ceramics with Oxynitride Additives, *J.Ceram.Process.Res.*, **1**, 12-9 (2000).

[13] V.A. Izhevskyi, L.A. Genova, A.H.A. Bressiani, and J.C. Bressiani, Liquid Phase Sintered SiC. Processing and Transformation Controlled Microstructure Tailoring, *Mat.Res.*, **3**(4), 131-38 (2000).

[14] V.A. Lavrenko, D.J. Baxter, A.D. Panasyuk, and M. Desmanion-Brut, High-Temperature Corrosion of AlN-Based Composite Ceramic in Air and in Combustion Products of Commercial Fuel. I. Corrosion of Ceramic Composites in the AlN-SiC System in Air and in Combustion Products of Kerosene and Diesel Fuel, *Powder Metallurgy and Metal Ceramics*, **43**(3-4), 179-86 (2004).

[15] D. Sciti, F. Winterhalter, and A. Bellosi, Oxidation Behaviour of Pressureless Sintered AlN-SiC Composite, *J.Mater.Sci.*, **39**, 6965-73 (2004).

[16] H.J. Choi, J.G. Lee, and Y.W. Kim, Oxidation Behaviour of Liquid-Phase Sintered Silicon Carbide with Aluminium Nitride and Rare-Earth Oxides (Re$_2$O$_3$, where Re=Y, Er, Yb), *J.Am.Ceram.Soc.*, **85**(9), 2281-86 (2002).

[17] Y.B. Pan, J.H. Qiu, and M. Morita, Oxidation and Microhardness of SiC-AlN Composite at High Temperature, *Mater.Sci.Bull.*, **33**(1), 133-39 (1998).

[18] S. Guo, N. Hirosaki, H. Tanaka, Y. Yamamoto, and T. Nakamura, Oxidation Behavior of Liquid-Phase Sintered SiC with AlN and Er$_2$O$_3$ Additives between 1200°C and 1400°C, *J.Eur.Ceram.Soc.*, **23**, 2023-29 (2003).

[19] K. Biswas, G. Rixecker, and F. Aldinger, Improved High Temperature Properties of SiC-Ceramics Sintered with Lu$_2$O$_3$-Containing Additives, *J.Eur.Ceram.Soc.*, **23**, 1099-1104 (2003).

[20] K. Biswas, G. Rixecker, and F. Aldinger, Effect of Rare-Earth Cation Additions on the High Temperature Oxidation Behaviour of LPS-SiC, *Mater.Sci.Eng. A*, **374**, 56-63 (2004).

[21] K.L. Luthra, Some New Perspective on Oxidation of Silicon Carbide and Silicon Nitride, *J.Am.Ceram.Soc.*, **74**(5), 1095-1103 (1991).

[22] H. Fritze, J. Jojic, T. Witke, C. Ruscher, S. Weber, S. Scherrer, R.Weiss, B. Schultrich, and G. Borchardt, Mullite Based Oxidation Protection for SiC-C/C Composites in Air at Temperatures up to 1900K, *J.Eur.Ceram.Soc.*, **18**, 2351-64 (1998).

[23] D. Cubicciotti, and K.H. Lau, Kinetics of Oxidation of Hot-Pressed Silicon Nitride Containing Magnesia, *J.Am.Ceram.Soc.*, **61**(11-12), 512-17 (1978).

CHARACTERIZATION OF THE RE-OXIDATION BEHAVIOR OF ANODE-SUPPORTED SOFCS

Manuel Ettler, Norbert H. Menzler, Hans Peter Buchkremer, Detlev Stöver
Jülich Forschungszentrum, Institute of Energy Research, IEF-1, Jülich, GERMANY

ABSTRACT

Today's state of the art substrate and anode material for anode-supported solid oxide fuel cells is a Ni-YSZ cermet. During operation or shut down accidental air break-in or controlled air feed on the anode side can result in re-oxidation of the metallic nickel. The volume expansion caused by Ni oxidation generates stresses within the substrate, the anode and the electrolyte. Those stresses can exceed the stability of the components, potentially promoting crack growth. This may lead to degradation of the SOFC or complete failure.

Therefore, cells manufactured at Jülich Forschungszentrum have been extensively characterized with respect to redox-stability. Investigations were carried out on samples consisting of three layers, the Ni-YSZ substrate, the Ni-YSZ anode and a YSZ electrolyte to study the influence of several re-oxidation parameters on the behaviour of the cell upon re-oxidation. This influence is to be referred to and explained by the fundamental processes that govern the progression of re-oxidation of substrate and anode of a SOFC. Failure criteria are defined with respect to mechanical integrity of the cell.

The experiments show, that mechanical integrity can be guaranteed by not exceeding certain values of the integral degree of oxidation of the cell. The limit value is strongly depending on the gradient in the local degree of oxidation in the substrate. Both integral degree of oxidation and gradient in the local degree of oxidation are determined by re-oxidation temperature, time of re-oxidation and incident air flow, as well as the substrate thickness. The progression of re-oxidation is governed by the two competing processes of Ni oxidation itself (determined mainly by the diffusion of Ni-ions through a NiO scale) and gaseous diffusion of oxygen molecules into the substrate and anode structure.

INTRODUCTION

The prevalent material for substrates and anodes in anode-supported solid oxide fuel cells (SOFCs) are porous composites of oxygen-ion conducting ceramics such as yttria-stabilised zirconia (YSZ) and Nickel. Cells based on such substrates and anodes have been found to show very good performance [1].

Typically the substrate is manufactured by a Coat-Mix® and warm pressing process or by tape casting followed by a pre-sintering step. The substrate is then coated with anode, electrolyte and cathode by vacuum slip casting or screen printing. Each coating is followed by heat treatment respectively. The raw materials for both substrate and anode are YSZ- and NiO-powders, but particle size distribution, composition and resulting microstructure may be different. Finally the NiO in the substrate and anode is reduced to Nickel during system start-up [1,2].

The presence of Nickel as catalytic component for the occurring chemical reactions and electronic conductor in the anode and substrate is essential, but also causes undesirable characteristics. One of the most important is the structural and dimensional instability if present Nickel is re-oxidized [1]. Under operating conditions fuel is supplied to the anode side of the cell and therefore the Nickel in the substrate and anode remains in the reduced state. However, if the fuel supply is interrupted oxygen is still able to pass from the cathode side through the electrolyte or via imperfect seals from outside of the system to the anode side of the cell. Interruption of the fuel supply may happen as a result of a fault in the system control or intentionally, e.g. upon system shut-down. The presence of oxygen on the anode side of the cell leads to re-oxidation of the Nickel which cannot be avoided. This can also occur, if the fuel utilisation is too high causing the oxygen activity to rise above that for equilibrium between Nickel and NiO [1].

RE-OXIDATION OF NICKEL IN SOFC SUBSTRATES AND ANODES

Nickel oxidation and reduction are well-investigated [3-13], but the results derived from these investigations cannot be applied offhand to Nickel based SOFC substrates and anodes because of their composite structure. Investigations of redox tolerance of SOFC substrates and anodes have gained further interest since SOFC systems are not only discussed as solutions for future stationary small-scale domestic, residential and industrial power generation, but also for transport and mobile applications, e.g. as auxiliary power units in planes, trucks or cars [14]. A mobile application of a SOFC system would necessitate the tolerance of a large number of intended redox cycles upon system shut-down during system lifetime. The effects of redox cycling on Nickel-based substrates and anodes for SOFCs have been studied by testing bars, discs and powders with the appropriate composition. Changes of weight and dimensions, mechanical integrity and microstructure were observed using techniques such as thermogravimetry, dilatometry and microscopy [15-21]. The influence of microstructure and other parameters on the redox behaviour and proposed potential solutions to the redox problem by adjusting various parameters were studied [19, 22-29]. The mechanism behind cell damage upon re-oxidation of the substrate and anode today is well known and has been described in various contributions in literature: Re-oxidized Nickel can be re-reduced, but the structure of the substrate and anode cannot be reverted to the original state [19-23, 25, 29]. The structural changes in the substrate and anode lead to microscopic and macroscopic dimensional changes that generate stresses in the substrate, anode and other cell components potentially promoting damages (e.g. cracks) in all layers of the cell and therefore might cause losses in cell performance or even complete failure of the cell [1, 21-23, 27].

MOTIVATION

Redox cycling causing changes in microstructure of the substrate and anode and significant volume changes does not necessarily lead to severe damages or complete failure of a cell. Some investigations on complete cells or half-cells (substrate-anode-electrolyte assemblies) have investigated the mechanisms that originate damages also in cell components other than substrate and anode potentially leading to complete disruption and failure of the cell. Substrate and anode expand upon re-oxidation, whereas electrolyte and cathode layers cannot follow this expansion. Tensile stresses arise in the cell. At first, these stresses can be relieved by bending of the cell. Once the tensile residual stresses in the electrolyte exceed tensile strength, cracks form in the electrolyte [16, 19, 22-24, 26, 30-33].

Stresses, curvature and cracking are strongly dependent on homogeneity of the oxidation and the degree of oxidation (DoO), defined as the ratio between the mass of oxygen absorbed by the substrate and the maximum mass of oxygen the substrate would have been able to absorb upon complete oxidation [16, 19, 22, 23, 31, 33]. Both, homogeneity and DoO, again strongly depend on the oxidizing conditions, e.g. temperature, time of oxidation and gas flow or cell characteristics like substrate thickness [31]. Malzbender et al. [23, 33] monitored the development of curvature of cells during redox cycling. The results showed an inhomogeneous re-oxidation at 800°C in air leading to complicated stress distributions ultimately resulting in electrolyte cracking. The purpose of this work is the characterisation of cells produced at Jülich Forschungszentrum with respect to redox stability. Investigations were carried out on half-cells to study the influence of re-oxidation temperature, time of re-oxidation, gas flow, substrate thickness and porosity on the behaviour of the cell upon re-oxidation. The study aims at connecting the redox behaviour with the homogeneity of oxidation and the degree of oxidation (DoO), defining failure criteria and referring the results to and explaining them by the fundamental processes that govern the progression of re-oxidation of substrate and anode of a SOFC.

EXPERIMENTAL

Various series of experiments were carried out on different types of samples. All samples consisted of three layers, the Ni-YSZ substrate, Ni-YSZ anode and YSZ electrolyte. The substrates were prepared by the Coat-Mix® process and warm pressing. The chemical composition is 56 wt.% NiO and 44 wt.% YSZ for both the Coat-Mix® substrate and the anode. Characterization of the Ni-YSZ-cermets with respect to redox tolerance was performed at free standing samples with dimensions 50×25 mm^2 and a thickness of 1.5, 1.0 or 0.5 mm to investigate the influence of the re-oxidation temperature, time of re-oxidation, incident air flow, substrate thickness and substrate porosity on the mechanical integrity of the substrate and other cell components. Examples of the free standing samples that were used for the tests are shown in Fig. 1.

Fig. 1: Example of free standing samples, dimensions: $50 \times 25 \times 1.5$ mm^3 and $50 \times 25 \times 0.5$ mm^3

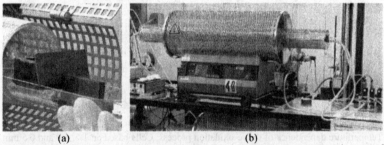

| (a) | (b) |

Fig. 2: Experimental setup; (a) sample on stand before the experiment, (b) furnace and setup for re-oxidation experiments

The samples for the tests are cut to their size out of big plates with a diamond saw and weighed in the as-prepared state. After that the NiO in the substrate and anode is completely reduced to metallic Ni by a standard procedure under an Ar/4%H$_2$ atmosphere and the samples are weighed again in the reduced state. Each sample is re-oxidized only once. Therefore the sample is placed onto a stand in a silica glass tube in the middle of a furnace. The experimental setup (see Fig. 2) is heated up to the re-oxidation temperature under a constant flow of Ar/4%H$_2$ through the glass tube.

When the re-oxidation temperature is reached the gas flow is switched to air. The flow rate is controlled by a mass flow controller. The air flow is maintained for the desired time of re-oxidation. Then the experimental setup is cooled down to room temperature under a constant flow of inert N$_2$ through the glass tube. Finally the sample is taken out of the setup and the experiment is evaluated.

Evaluation includes weighing the sample again in the re-oxidized state allowing the determination of the integral degree of oxidation (DoO) by the following relation:

$$DoO = \frac{m(reoxidised)}{m(as-prepared) - m(reduced)} \cdot 100 \tag{1}$$

Substrate and electrolyte are checked for cracks via microscope, eventually followed by further microscopic or SEM investigations of areas of fracture or cross sections.

In every set of experiments, consisting of various tests, the same cell type is used. Only one parameter is changed from one test to the next while all other parameters are fixed. To compare the results of the different sets of experiments they are represented in graphic form. The graphs are shown and interpreted and the results are discussed in the following section.

RESULTS

Experiments were carried out to identify the governing processes of re-oxidation of the substrates and to define failure criteria with respect to re-oxidation parameters. The latter can only apply to intended re-oxidation, e.g. upon system shut-down. They are defined with respect to mechanical integrity. Electrochemical performance is not subject to investigation in this approach. Re-oxidation upon system failure, for example via air break-in through imperfect seals, cannot be controlled and will therefore damage the cell in the long run, independent of the re-oxidation conditions.

The results of three sets of experiments with half cells on the basis of 1.5, 1.0 and 0.5 mm thick Coat-Mix$^{\circledR}$ substrates, re-oxidized for 15 minutes with an air flow rate of 1.2 l/min, show that the kinetics of the re-oxidation process is strongly determined by the temperature.

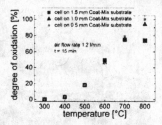

Fig. 3: Temperature dependence of the re-oxidation process: Cells based on 1.5, 1.0 and 0.5 mm Coat-Mix$^{\circledR}$ substrates re-oxidized at temperatures between 300 and 800°C for 15 min with 1.2 l/min air flow

In every set the temperature is varied between 300 and 800°C (see Fig. 3). No re-oxidation takes place at 300°C. At 400°C the kinetics of the reaction is very slow, a DoO of around 5% is reached for all three cell types not leading to any damage. Also at 500°C all cell types show the same DoO (almost 20%). The cells are not damaged, although the kinetics of the re-oxidation is much faster. A slightly different DoO of the three cell types is observed at 600°C. However, the difference is insignificant. For all three substrates the DoO reaches a value of around 45%. Substrate and electrolyte show no damages. Cracks in the electrolyte appear after the experiments at 700°C. The DoO is still very similar for all cell types with values of around 78%. At 800°C the DoO for the 1.5 mm substrate is about the same as in the experiment at 700°C, while the thinner substrates have significantly higher

DoOs. The 0.5 mm substrate is fully re-oxidized. The 1 mm substrate is re-oxidized to 95%. The electrolytes of all cells are completely destroyed.

The evaluations of the various tests show that the re-oxidation is less detrimental to the cells at temperatures lower than 600°C because the DoOs that can be reached in an intended redox cycle with an appropriate time and air flow are rather low. Re-oxidation at higher temperatures may be damaging to the cells. An explanation for the different DoOs at 800°C depending on the substrate thickness will be given in the discussion section. The significant influence of temperature on the re-oxidation process can also be demonstrated with the following graphs of Fig. 4 and Fig. 5. They show an overview of various series of experiments at temperatures between 400 and 800°C with cells based on 1.5 and 1.0 mm Coat-Mix[*] substrates. The graphs show the DoO versus time of re-oxidation at a constant air flow and air flow rate at a constant time of re-oxidation for different temperatures. The higher the temperature. the faster the DoO rises. At 400°C the re-oxidation is extremely slow. The DoO is not much higher after 120 minutes than after 15 minutes. At 500 and 600°C the kinetics gets faster but is still slow compared to 700 and 800°C. The curves at 700 and 800°C differ only slightly for the 1 mm substrate and are nearly congruent for the 1.5 mm substrate.

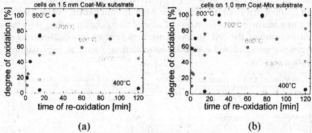

(a)   (b)

Fig. 4: Temperature dependence of the re-oxidation process: Series of experiments on cells on Coat-Mix[®] substrates re-oxidized at temperatures between 400 and 800°C with 1.2 l/min air flow and varying time of re-oxidation. (a) 1.5 mm substrates, (b) 1.0 mm substrates

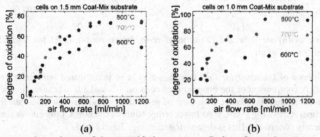

(a)   (b)

Fig. 5: Temperature dependence of the re-oxidation process: Series of experiments on cells based on Coat-Mix[®] substrates re-oxidized at temperatures between 600 and 800°C for 15 min with varying air flow. (a) 1.5 mm substrates, (b) 1.0 mm substrates

The series of experiments allowed the definition of maximum tolerable values of the DoO depending on temperature and substrate thickness as failure criteria for Coat-Mix[®] substrates. The limit values depend only on temperature and cell type. How the DoO is reached, in a short time with high flow or vice versa, does not play a role. For the 1.5 mm substrate first mechanical damage was observed at DoOs of 60% for 600°C, 40% for 700°C and 20% at 800°C. The 1 mm substrate did not show any damages even after complete re-oxidation at 600°C. The limits for 700 and 800°C were DoOs of 55 and 20%. Finally the limit values for 0.5 mm substrates were DoOs of about 70% at 600°C, about 60% at 700°C and about 30% at 800°C. A proposal for an explanation of the strong temperature dependence of the limit values will be given in the discussion section.

The failure criteria are defined by maximum tolerable temperature dependent DoO values. As stated earlier the limits are independent of the time of re-oxidation and air flow rate. However, these parameters may be important for system operation. In intended redox cycles upon system shut-down a well defined volume of air will be applied to the anode side of the cell until the system is cooled down. If air flow rate and time of re-oxidation can be controlled together with temperature, the DoO can be minimized. This shows Fig. 6. In two sets of experiments at 600 and 800°C samples based on 1.5 mm Coat-Mix$^®$ substrates were re-oxidized with a total air volume of 18 l. The air volume was applied to the cells in various combinations of air flow rate and time of re-oxidation. The results show, that it is beneficial to apply the air in the shortest possible time with the highest possible flow. At 800°C the complete re-oxidation after 60 min with a flow of 300 ml/min could be reduced to less than 75% by applying the volume in 15 min with a flow of 1.2 l/min. At 600°C re-oxidation for 120 min and a flow rate of 150 ml/min resulted in a DoO of about 80%. After 15 min with 1.2 l/min the DoO stayed under 30%. So the choice of a reasonable combination of time of re-oxidation and air flow rate can be crucial for the mechanical integrity of the cell. In this sense both time of re-oxidation and air flow rate are important parameters for intended and controlled re-oxidation. Again an explanation for this will be given in the discussion section.

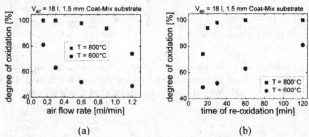

(a)   (b)

Fig. 6: Influence of (a) air flow and (b) time of re-oxidation on the DoO for re-oxidation with a constant volume of air

The influence of the substrate thickness has also been investigated with cells based on Coat-Mix$^®$ substrates. A comparison of the behaviour of cells on 1.5 and 1.0 mm substrates shows that the DoO of thinner substrates rises faster with time of re-oxidation and air flow rate than that of thicker ones at high temperatures (T ≥ 700°C). At lower temperatures (T ≤ 600°C) the curves for the different substrates are nearly congruent. This is demonstrated in Fig. 7 and Fig. 8.

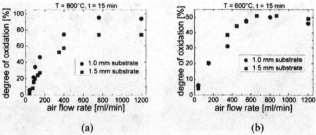

(a)   (b)

Fig. 7: Influence of the substrate thickness: DoO vs. air flow rate for a constant time of re-oxidation of 15 min, comparison of the behaviour of cells on the basis of 1.5 and 1.0 mm Coat-Mix® substrates; (a) T = 800°C, (b) T = 600°C

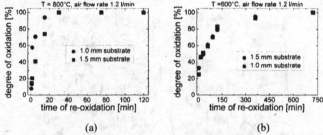

(a)   (b)

Fig. 8: Influence of the substrate thickness: DoO vs. time of re-oxidation for a constant air flow rate of 1.2 l/min, comparison of the behaviour of cells on the basis of 1.5 and 1.0 mm Coat-Mix® substrates; (a) T = 800°C, (b) T = 600°C

The graphs show a comparison of the DoO versus air flow at a constant time of re-oxidation and time of re-oxidation at a constant air flow rate for cells based on 1.5 and 1.0 mm Coat-Mix® substrates at 800 and 600°C. The curves at 800°C take a significantly different course, whereas no difference can be observed at 600°C.

To explain the various effects of temperature, time of re-oxidation, air flow and substrate thickness additional microscopic and SEM investigations have been carried out on re-oxidized samples. Microscopic pictures of areas of fracture of such cells revealed that a big gradient in the local degree of oxidation appears at higher temperatures (T ≥ 700°C). At lower temperatures no such gradient can be observed, the substrate is re-oxidized homogenous (see Fig. 9). Half of the substrate of the sample re-oxidized at 800°C is completely re-oxidized (local DoO 100%), the other half is still completely reduced (local DoO 0%). A re-oxidation front appears in the substrate that separates the re-oxidized part from the reduced part. The sample in Fig. 9 (b) re-oxidised at 600°C has a local DoO of about 50% everywhere in the substrate, no re-oxidation front can be observed, no gradient appears in the local DoO.

SEM pictures of cross sections of the two parts confirm these observations. The area of the substrates close to the surface is fully re-oxidized, whereas the area close to the anode is completely reduced (Fig. 10). This gets even clearer in comparison to according SEM pictures of a completely re-oxidized sample (Fig. 11). Pores appear black in the SEM pictures, the dark grey phase is NiO, the light grey phase is YSZ or Ni, as the contrast between the two cannot be resolved in the SEM.

(a)   (b)

Fig. 9: Microscopic pictures of areas of fracture of two cells on a 1.5 mm Coat-Mix® substrate, both re-oxidized to 50%; (a) re-oxidation at 800°C, gradient in local DoO, (b) re-oxidation at 600°C, homogenous re-oxidation

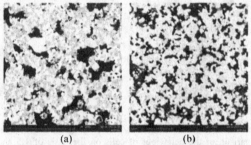

(a)   (b)

Fig. 10: SEM pictures of cross sections of a partially re-oxidized sample; (a) area close to the surface (local DoO 100%), (b) area close to the anode (local DoO 0%)

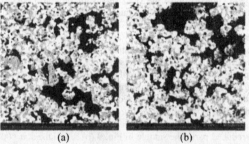

(a)   (b)

Fig. 11: SEM pictures of cross sections of a fully re-oxidized sample; (a) area close to the surface (local DoO 100%), (b) area close to the anode (local DoO 100%)

Microscopic pictures of areas of fracture were taken of samples out of each series of experiments systematically. They were analysed with the Analysis program to measure the fraction of the substrate that is re-oxidized. The results for the re-oxidation experiments on half cells on the basis of 1.5 and 1.0 mm Coat-Mix® substrates with variable air flow for 15 min at 800,700 and 600°C are shown in Fig. 12:

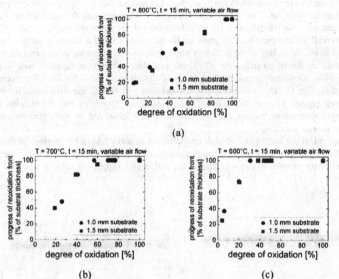

(a)

(b)   (c)

Fig. 12: Progress of re-oxidation front vs. degree of oxidation in 1.5 and 1.0 mm Coat-Mix® substrates upon re-oxidation for 15 min with variable air flows at different temperatures; (a) T = 800°C, (b) T = 700°C, (c) T = 600°C

All curves show a small offset that is due to the fact, that the border between the re-oxidized and reduced area is not a sharp line. In the analysis the border line was set to the beginning of the reduced area. A part of the re-oxidized area is not completely re-oxidized, the measured integral DoO is a little bit lower than it would be if the border was really a sharp line. The graphs show a linear relation between DoO and progress of the re-oxidation front. The slope is temperature dependent, it increases with decreasing temperature. Obviously the progress of the re-oxidation front does not depend on the substrate thickness. The graphs will be further interpreted in the discussion section.

DISCUSSION

The progress of the re-oxidation in the substrate is governed by two fundamental processes. On the one hand the oxidation of the metallic Ni particles in the substrate determined by the growth of the oxide scale via outward Ni diffusion determines the kinetics of re-oxidation. This process shows a parabolic or sub-parabolic behaviour with respect to time and is strongly temperature dependent [3-13]. The influence of the process can explain the fundamental form of the curves of DoO versus time of re-oxidation typically found in our investigations. On the other hand the process is superposed by the gaseous diffusion of oxygen molecules into the substrate structure. It determines in which parts of the substrate the Ni particles can be re-oxidised and therefore can cause the creation of a re-oxidation front. It is dependent on time, air flow and substrate thickness. In the sense that the second process is responsible for oxygen supply and the first process for oxygen consumption, they are competing and influence each other. The relation of the fundamental processes to each other, i.e. which of the two controls the kinetics of re-oxidation, determines the progress of re-oxidation and therefore the behaviour of the cell upon re-oxidation.

The progress of the re-oxidation front (see Fig. 12) reveals that the diffusion of oxygen into the substrate structure controls the re-oxidation kinetics at 800°C. Oxygen that diffuses into the substrate is consumed immediately for oxidation of Ni particles close to the substrate surface. Only when these particles are re-oxidized completely other particles further away from the substrate surface begin to be re-oxidized and so on. So at 800°C the diffusion of oxygen into the substrate is slower than the growth of the oxide scale on the Ni particles. At 700°C the complete substrate is supplied with oxygen at a DoO of about 50%. Obviously oxide scale growth is slower than the diffusion of oxygen into the substrate and therefore is rate controlling for re-oxidation. At 600°C the difference between both processes is even bigger. The identification of the fundamental processes and the insights discussed before are very important to understand the various effects observed in our experiments and the according behaviour of the cells.

The gradient in the local DoO at high temperatures induces additional stresses in the layers of the cell. The stresses are first minimized through warpage of the cells. The gradient in the local DoO is bigger at higher temperatures, so the warpage of the cell is also bigger for cells re-oxidized at higher temperatures (see Fig. 13).

(a)   (b)

Fig. 13: Two cells on the basis of a 1.5 mm Coat-Mix[®] substrate with a DoO of about 50%;   (a) T = 800°C, cell shows distinct warpage. (b) T = 600°C, cell remains flat

When the stresses exceed a given value cracks arise in the electrolyte. So the gradient in the local DoO is also responsible for the temperature dependence of the maximum tolerable values of the integral DoO defined in the results section. The higher the re-oxidation temperature is, the bigger the gradient in the local DoO and the greater the additional stresses are. So it is logical that the limits for the DoO for mechanical integrity are lower for higher temperatures and vice versa.

The influence of the substrate thickness on the DoO at 800°C can be explained by the different fractions of the substrate that are supplied with oxygen. The thinner the substrate is, the bigger the fraction of the substrate that is supplied with oxygen and the higher the integral DoO. At lower temperatures (T ≤ 600°C) the growth of the oxide scale on the Ni particles is the rate controlling process. So at these temperatures the influence of substrate thickness is less distinct (compare Figs. 3, 7 and 8).

Also the nearly congruent curves for the re-oxidation of 1.5 mm Coat-Mix[®] substrates at 700 and 800°C (see Figs. 4 (a) and 5 (a)) can be explained by the relation between the two fundamental processes. At 800°C the oxidation kinetics is faster, but a smaller fraction of the substrate is supplied with oxygen. So the influences of both processes neutralize each other. The 1 mm substrate seems to be supplied with oxygen almost completely after a relatively short time, so the difference in oxidation kinetics between 700 and 800°C actually appears in the curves shown in the result section (see Figs. 4 (b) and 5 (b)).

The fact that the DoO can be minimized by choosing the minimum possible time of re-oxidation with the maximum possible air flow at re-oxidation with a given volume of air can also be explained by the fundamental processes. Both are time dependent and therefore a shorter re-oxidation time leads to a smaller DoO via minor progression of the re-oxidation front at 800°C and a minor growth of the oxide scale on the Ni particles at 600°C.

SUMMARY

An extensive characterization of the behaviour upon re-oxidation with respect to mechanical integrity has been carried out to investigate the redox stability of various types of SOFCs. The substrates of these cell types originate from a Coat-Mix® and warm pressing process. The influences of the re-oxidation parameters temperature, time of re-oxidation, air flow rate and substrate thickness have been studied. Failure criteria were defined for every cell type by determining maximum tolerable values for the degree of oxidation. By analyzing the various series of experiments and with the help of additional microscopic investigations on re-oxidized samples the progress of the re-oxidation was referred to two competing fundamental processes, the growth of the oxide scale on the Ni particles in the substrate and the diffusion of oxygen into the substrate. All effects that have occurred in the various observations could be explained with recourse to these fundamental and rate determining processes.

ACKNOWLEDGEMENT

The work is part of the BMWi-financed (Federal Ministry of Economics) project "Entwicklung Nebenaggregate SOFC-APU" (ENSA).

REFERENCES

1) D. Sarantaridis and A. Atkinson, Redox Cycling of Ni-Based Solid Oxide Fuel Cell Anodes: A Review, Fuel Cells, 7 [3], p. 246 (2007)

2) W.A. Meulenberg, N.H. Menzler, H.-P. Buchkremer and D. Stöver, Manufacturing Routes and State-of-the-art of the Planar Juelich Anode-Supported Concept for Solid Oxide Fuel Cells, Ceramic Transactions, 127, p. 99 (2002)

3) M. Radovic and E. Lara-Curzio, Mechanical properties of tape cast nickel-based anode materials for solid oxide fuel cells before and after reduction in hydrogen, Acta Mater., 52 (20), p. 5747 (2004)

4) J.T. Richardson, R. Scates and M.V. Twigg, X-ray diffraction study of nickel oxide reduction by hydrogen, Appl. Catal. A, 246 (1), p. 137 (2003)

5) T.A. Utigard, M. Wu, G. Plascencia and T. Marin, Reduction kinetics of Goro nickel oxide using hydrogen, Chem. Eng. Sci., 60 (7), p. 2061 (2005)

6) A. Atkinson, Growth of NiO and SiO_2 thin films, Philos. Mag. B, 55 (6), p. 637 (1987)

7) A. Atkinson, Transport processes during the growth of oxide films at elevated temperature, Rev. Mod. Phys., 57 (2), p. 437 (1985)

8) A. Atkinson, R. I. Taylor and A.E. Hughes, A quantitative demonstration of the grain boundary diffusion mechanism for the oxidation of metals, Philos. Mag. A, 45 (5), p. 823 (1982)

9) A. Atkinson, R. I. Taylor and P. D. Goode, Transport processes in the oxidation of Ni studied using tracers in growing NiO scales, Oxid. Met., 13 (6), p. 519 (1979)

10) A. Atkinson and R. I. Taylor, The diffusion of Ni in the bulk and along dislocations in NiO single crystals, Philos. Mag. A, 39, p. 581 (1979)

11) A. Atkinson and D.W. Smart, Transport of Nickel and Oxygen during the Oxidation of Nickel and Dilute Nickel/Chromium Alloy, J. Electrochem. Soc., 135 (11), p. 2886 (1988)

12) R. Karmhag, G.A. Niklasson and M. Nygren, Oxidation kinetics of large nickel particles, J. Mater. Res., 14 (7), p. 3051 (1999)

13) R. Karmhag, G.A. Niklasson and M. Nygren, Oxidation kinetics of small nickel particles, J. Appl. Phys., 85 (2), p. 1186 (1999)

14) L. Blum, H.-P. Buchkremer, S. Gross, A. Gubner, L.G.J. (Bert) de Haart, H. Nabielek, W.J. Quadakkers, U. Reisgen, M.J. Smith, R. Steinberger-Wilckens, R.W. Steinbrech, F. Tietz and I.C. Vinke, Solid Oxide Fuel Cell Development at Forschungszentrum Juelich, Fuel Cells, 7 [3], p. 204 (2007)

15) N.M. Tikekar, T.J. Armstrong and A.V. Virkar, Reduction and Re-oxidation Kinetics of Nickel-Based Solid Oxide Fuel Cell Anodes, Proc. Solid Oxide Fuel Cells VIII, (Eds. S.C. Singhal, M. Dokiya), Paris, France, p. 670 (2003)

16) G. Stathis, D. Simwonis, F. Tietz, A. Moropoulou and A. Naoumides, Oxidation and resulting mechanical properties of $Ni/8Y_2O_3$-stabilized zirconia anode substrate for solid-oxide fuel cells, J. Mater. Res., 17 (5), p. 951 (2002)

17) S. Modena, S. Ceschini, A. Tomasi, D. Montinaro and V.M.Sglavo, Reudction and Re-oxidation Processes of NiO/YSZ Composite for Solid Oxide Fuel Cell Anodes, J. Fuel Cell Sci. Tech., 3, p. 487 (2006)

18) A.C. Mueller, D. Herbstritt, A. Weber and E. Ivers-Tiffée, , Proc. 4th European SOFC Forum, (Ed. J.A. McEvoy), Lucerne, Switzerland, p. 579 (2000)

19) D. Waldbillig, A. Wood and D. G. Ivey, Thermal analysis of the cyclic reduction and oxidation behaviour of SOFC anodes, Solid State Ionics, 176, p. 847 (2005)

20) T. Klemensø, C. C. Appel and M. Mogensen, In Situ Observations of Microstructural Changes in SOFC Anodes during Redox Cycling, Electrochem. Solid-State Lett., 9, p. A403 (2006)

21) T. Klemensø, C. Chung, P. H. Larsen and M. Mogensen, The mechanism behind redox instability of anodes in high-temperature SOFCs, J. Electrochem. Soc., 152 (11), p. A2186 (2005)

22) D. Waldbillig, A. Wood and D. G. Ivey, Electrochemical and microstructural characterization of the redox tolerance of solid oxide fuel cell anodes, J. Power Sources, 145, p. 206 (2005)

23) J. Malzbender, E. Wessel and R.W. Steinbrech, Reduction and re-oxidation of anodes for solid oxide fuel cells, Solid State Ionics, 176, p. 2201 (2005)

24) M. Cassidy, G. Lindsay and K. Kendall, The reduction of nickel-zirconia cermet anodes and the effects on supported thin electrolytes, J. Power Sources, 61, p. 189 (1996)

25) D. Fouquet, A. C. Müller, A. Weber and E. Ivers-Tiffée, Kinetics of oxidation and reduction of Ni/YSZ cermets, Proc. 5th European SOFC Forum, (Ed. J.Huijsmans), Lucerne, Switzerland, p. 467 (2002)

26) G. Robert, A. Kaiser and E. Batawi, Anode Substrate Design for RedOx-Stable ASE Cells, Proc. 6th European SOFC Forum, (Ed. M. Mogensen), Lucerne, Switzerland, p. 193 (2004)

27) A. Wood, M. Pastula, D. Waldbillig and D.G. Ivey, Initial Testing of Solutions to Redox Problems with Anode-Supported SOFC, J. Electrochem. Soc., 153 (10), p. A1929 (2006)

28) L. Grahl-Madsen, P.H.Larsen, N. Bonanos, J. Engell and S. Linderoth, , Proc. 5th European SOFC Forum, (Ed. J.Huijsmans), Lucerne, Switzerland, p. 82 (2002)

29) Y. Zhang, B. Liu, B. Tu, Y. Dong and M. Cheng, Redox cycling of Ni-YSZ anode investigated by TPR technique, Solid State Ionics, 176, p. 2193 (2005)

30) W. Li, K. Hasinska, M. Seabaugh, S. Swartz and J. Lannutti, Curvature in solid oxide fuel cells, J. Power Sources, 138, P. 145 (2004)

31) M. Ettler, G. Blaß and N.H. Menzler, Characterisation of Ni–YSZ-Cermets with Respect to Redox Stability, Fuel Cells, 7 [5], p. 349 (2007)

32) G. Robert, A. Kaiser, K. Honegger and E. Batawi, Anode Supported Solid Oxide Fuel Cells with a Thick Anode Substrate, Proc. 5th European SOFC Forum, (Ed. J. Huijsmans), Lucerne, Switzerland, p. 116 (2002)

33) J. Malzbender, E. Wessel, R.W. Steinbrech and L. Singheiser, Reduction and re-oxidation of anodes for solid oxide fuel cells, Proc. 28th International Conference on Advanced Ceramics and Composites: A, (Eds. E. Lara-Curzio, M.J. Readey), Cocoa Beach, USA, p. 387 (2005)

HEALING BEHAVIOR OF MACHINING CRACKS IN OXIDE-BASED COMPOSITE CONTAINING SiC PARTICLES

Toshio Osada
Post-graduate student, Department of Energy and Safety Engineering
79-5 Tokiwadai, Hodogaya-ku, Yokohama, 240-8501, Japan

Wataru Nakao, Koji Takahashi, Kotoji Ando
Department of Energy and Safety Engineering
79-5 Tokiwadai, Hodogaya-ku, Yokohama, 240-8501, Japan

ABSTRACT

Machined mullite containing 15 vol.% SiC particles has been subjected to various heat-treatments. The effect on fracture stress has been investigated as a function of crack-healing temperature and time. By heating at 1673 K for 10 h prior to monotonic bending and static fatigue testing, specimens fractured from sites other than the machining cracks. Thus, it was concluded that the machining cracks were completely healed by this heat treatment. From these results, the monotonic and static fatigue strength at elevated temperature of the healed sites exhibited the same levels as base material (mullite/ SiC composite). It is proposed that the crack-healing treatment is a viable economic technique to be applied to structural ceramics.

INTRODUCTION

Ceramics have been used in various fields as machined components. In particular, oxide-based ceramics is expected to be employed in structural components operating at high temperatures in various atmospheres, e.g., automotive engine, because of its excellent mechanical properties and oxidation resistance. However, the heavy machining introduces easily numerous non-acceptable damages, such as cracks, chips and damaged layers, because oxide-based ceramics has high brittleness and high hardness [1]. Especially the cracks in the stress concentrated part and component, such as blade stud, reduce the structural integrity critically. General post machining operations, i.e. polishing and lapping together with nondestructive inspection, are performed to remove the non-acceptable flaws. Although these techniques are very costly, these cannot eliminate minute flaws. Thus, the reliability of the ceramic components cannot be secured by these techniques.

Whether this function can also be useful for cracks caused by machining is a very interesting research subject. If such cracks can be healed and crack-healed zones exhibits the same fatigue strength as the base materials, ceramic components can offer improved reliability without costly final machining. Many investigations [2-20] on repair for crack by heat treatment were reported on oxide-based ceramics or oxide-based composite containing SiC. The reported mechanism was based on (a) re-sintering of matrix or (b) oxidation of SiC. Matsuo et al. [6] reported that the crack tip in the machined monolithic alumina was re-sintered by the action of compressive residual stress and heat. However, the strength recovery was not complete. Niihara et al. [7] reported the same strength recoveries in alumina/ 5 vol.% SiC nanocomposite as monolithic alumina. On the other hand, Ando et al. [8-13] and Wu et al. [14] reported the large strength recovery could be realized by the crack-healing due to the oxidation of SiC at high-temperature in air as following chemical reaction,

$$SiC + 3/2\ O_2 = SiO_2 + CO + 943\ kJ/mol \qquad (1)$$

Moreover, Ando *et al.* proposed three conditions for complete strength recovery by the oxidation of SiC as follows; (I) Healing material (SiO_2) should exhibit the same level or higher strength as base material, (II) SiO_2 should be bonded to base material strongly, (III) SiO_2 should fill the crack completely. To achieve the above (I) and (II) conditions, large heat generation of 943 kJ/mol is very important. Alternatively to achieve above (III) condition, more than 10 vol% SiC is required. If above three conditions were satisfied completely, most samples fractured outside the crack-healed zone

However, Ando *et al.* have investigated the healing of not the machining cracks but the indented crack in the oxide-based ceramics, i.e., mullite [11-13, 18], and alumina [8-9, 15-17] (based materials), reinforced by SiC particles or SiC whiskers. Moreover, present authors [19-20] have investigated alumina /SiC composite could heal the machining cracks completely. Thus, investigating the self-healing of the machining cracks in the mullite containing SiC is necessary to actualize the advanced machining in which one can substitute the crack-healing process for the post machining operations.

In this study, mullite/ 15 vol.% SiC particles composite was sintered and the crack-healing behavior for the cracks caused by machining was investigated. Small bending specimens ($22 \times 4 \times 3$ mm) were made. These specimens were machined into semi-circular groove so that they have stress concentration factor to replicate that of components. The machined specimens were crack-healed under various conditions. The fracture stress and static fatigue strength of these specimens after crack-healing were evaluated systematically. From the obtained results, the crack-healing effect on the cracks was discussed.

EXPERIMENTAL
Material, specimen preparation and crack-healing conditions

The mullite powder (KM101, kioritzz Co. Ltd., Nagoya, Japan) used in this study has an average particle size of 0.76 μm. The SiC powder (Ultrafine grade, Ibiden Co. Ltd., Ogaki, Japan) used has an average particle size of 0.27 μm. For mixing the raw powder, the mixture of alumina powders and 15 vol.% SiC powders were blended well in alcohol for 24 h. Rectangular plates ($90 \times 90 \times 6$ mm) were hot pressed in N_2 at 1923 K under 35 MPa for 2 h The sintered plates were cut into $3 \times 4 \times 22$ mm rectangular bar specimens as shown in figure 1. The specimens were polished to a mirror finish on one face according to the Japan Industrial Standard (JIS) [21]. The edges of the specimens were beveled 45° to prevent fractures due to edge-cracks. In this paper these specimens were called "smooth specimens". As shown in Figure 1, semicircular groove was made at the center of the smooth specimens by using diamond-coated grinding wheel ($\phi 200$). The other grinding conditions are listed in Table I, where, maximum depth of cut by one pass (d) was 20 μm.

The machined specimens were subjected to crack-healing treatment by heat-treating in air at 1373 - 1673 K for 1 or 10 h. The specimens are here called as machined specimen healed. Also the smooth specimens were crack-healed at 1573 K for 1 h in air. These specimens are defined as healed smooth specimen. Even the smooth specimen involved minute flaws such as surface-crack and surface-pore. The flaws can be completely healed by heat-treatment [10-13]. Thus, all the fractures of the healed smooth specimens initiate from largest embedded flaw such as pore.

In this paper, therefore, the healed smooth specimens were treated as specimens without surface flaws.

Fatigue and fracture tests

The monotonic bending tests and static fatigue tests were performed on a three-point bending system with a span of 16 mm as shown in figure 1. The crosshead speed of the monotonic bending tests was 0.5 mm/min. The tests were performed at room and elevated temperatures. From the bending moment measured as the specimens fractured, M_F, the following equation was used to evaluate the nominal fracture stress (σ_{NF}) of the machined specimens.

$$\sigma_{NF} = \frac{M_F}{Z} \tag{2}$$

where Z is the section modulus of the machined specimens. The values of Z for the semicircular groove was 4.2 mm³.

Using stress concentration factor, K_t, and the following equation, one can evaluate the fracture stress (σ_F) at the stress concentration feature:

$$\sigma_F = K_t \sigma_{NF} \tag{3}$$

The K_t values relating to the semicircular groove, 1.2 was obtained from a data handbook [21].

Static fatigue tests were performed in air at elevated temperatures of 1373 K using a hydraulically controlled testing machine with an electric furnace and terminated in 100 h according to the Japan Industrial Standard (JIS R1632) [23]. The applied stress (σ_{app}) at which the specimen did not fracture within 3.6×10^5 sec (100 h) was defined as the static fatigue limit (σ_{t0}) in this study.

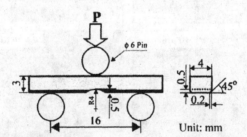

Figure 1. Schematic illustration of three-point loading system and test specimen size and machining figuration

Table I. Grinding conditions

Specimen preparation	Semicircular groove (R4)
Grindstone and dressing condition	#230 diamond metal-bond
Number of rotations(rpm)	3000
Table feed speed (mm/min)	15000
Cut depth by one pass (μm/pass)	3,5,10,15 and 20

RESULTS AND DISCUSSION

Typical machining damages

Figures 2 (a) and (b) show SEM images showing typical machining damages of as-machined specimens. As shown in the figures, it was found that numerous damages, such as a crack and a chip, introduced at the machined surface by heavily machining. Figures 2 (a) show SEM images of the machining crack in the as-machined specimen ($d = 10$ μm/pass). As shown in the figures 2 (a), the fracture origin was confirmed to be cracks introduced by machining. This fact suggested that the machining crack should be the severest flaw of all the machining damages and the embedded flaws. The major crack was likely formed by connecting the semi-elliptical cracks, thereby having wavy tip. Moreover, the maximum depth of this crack was found to be approximately 40 μm. As a result, the fracture origin of as-machined specimens was found to be such a large machining crack. On the other hand, it was found that cleavages, i.e., a grinding track, a grain pull-out and a chip, were introduced at the machined surface, resulting in rough surface as shown in figure 2 (b). Moreover, cleavages size was found to increase with increasing cut depth by one pass as shown in figure 2 (b-1) ~ (b-3). Therefore, it was concluded that numerous damages, such as cracks and chips, introduced at the machined surface by heavily machining.

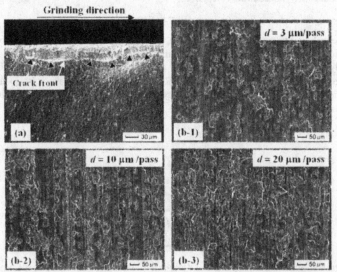

Figure 2. SEM images showing (a) machining crack and (b) machined surface of as-machined specimen: (b-1) $d = 3$ μm/pass, (b-2) $d = 10$ μm/pass and (b-3) $d = 20$ μm/pass

Crack-healing behavior

Figure 3 shows the effect of the crack-healing conditions on the strength recovery behavior of the machined specimen. The cut depth by one pass, d was 10 μm. The symbols $\Diamond$ in left column and ● in right column show the fracture stress (σ_F) of as-machined specimen and healed smooth specimen, respectively. The average σ_F of these specimens were found to be 177 MPa and 628 MPa, respectively, indicative of an 100% strength recovery upon crack-healing.

The symbols ◆ and □ in figure 3 show the σ_F of machined specimen healed at elevated temperatures for 1 or 10 h, respectively. The σ_F of healed specimens increases gradually up to 1573 K, above which it increases abruptly. The average σ_F has a maximum at healing temperature of 1673 K for 10 h. Moreover, the σ_F was almost equal to the fracture stress of healed smooth specimens. From these results, the optimal crack-healing conditions for machining cracks was defined as heating at 1673 K for 10 h. However, it was found that these conditions are quite different from the optimized crack-healing conditions for indention cracks within the same material which was reported previously as 1573 K for 1 h [10-13]. Two explanations for this difference are raised, (1) the difference in the state of the subsurface residual stress associated with the different crack geometries and (2) oxidation of SiC by the heat generation during machining. The machining crack was possibly closed by the action of the compressive residual stress which would result in a reduction in the supply of oxygen to the crack surfaces. Moreover, before the crack-healing treatment, if SiC particles were already covered with a thin oxidation layer due to the heat generated during grinding, this would lead to the decrease in the oxidation rate of SiC particles.

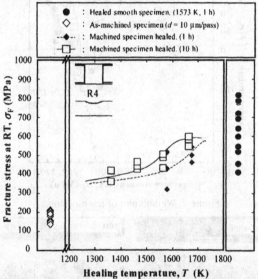

Figure 3. Effect of crack-healing temperature and time on strength recovery of machined specimen.

Statistical analysis of fracture stress

Figure 4 shows the Weibull plots of the fracture stresses of the as-machined specimen (◇) and the machined specimen healed at 1673 K for 10 h (□). Where $d = 10$ μm/pass. Also the Weibull plot of σ_F of the healed smooth specimen (●) was shown in this figure. A two-parameter Weibull function is given by

$$F(\sigma_F) = 1 - \exp\left\{ -\left(\frac{\sigma_B}{b}\right)^c \right\}$$
(4)

where b is a scale parameter, c is a shape parameter.

By crack-healing treatment at 1673 K for 10 h, the values of b of the machined specimen increased considerably from 189 MPa to 582 MPa, with this value being almost equal to that of healed smooth specimen. This means that machined specimens healed recovered its strength almost completely due to crack-healing.

Furthermore, it was found that the machining cracks were completely healed by crack-healing as shown in figure 5 (a). Figures 5 (a-1) and (a-2) show the SEM images of fracture origin and detail of point A, respectively of the machined specimen healed at 1673 K for 10 h. As shown in these figures, all machined specimen healed fractured from sites other than the

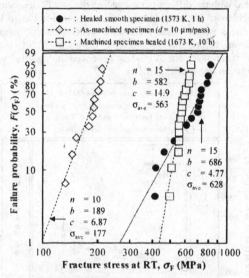

Figure 4. Weibull plot of fracture stress

Figure 5. SEM images of fracture origin of machined specimen healed at 1673 K for 10 h: (a-1) fracture origin and (a-2) detail of point A.

crack-healed zone, i.e., chipping at the surface. It could be clearly understood by un-reacted SiC agglomerates, which show the fracture origin was not oxidation products. From this result, it could be concluded that numerous machining cracks were completely healed by the optimized crack-healing treatment, and these chipping could not be healed, resulting in slightly decreased strength. Therefore, the crack-healing treatment is an effective technique for increasing the reliability of machined mullite/ SiC composite ceramics.

Effect of cut depth by one pass on fracture stress of machined specimen.

Figure 6 shows the effect of depth of cut by one pass (d) on the σ_F of healed machined specimens containing a semi-circular groove. Also the data on as-machined specimens were included in figure 6. The symbol □ shows the σ_F of machined specimen healed at 1673 K for 10 h. Also the symbol ◇ shows the σ_F of as-machined specimen. The symbol ● in the left column shows the σ_F of the healed smooth specimen.

Throughout the whole range of the d, a considerably strength recovery was attained by crack-healing treatment at 1673 K for 10 h, because these average strengths were almost equal to that of healed smooth specimen. Thus, crack-healing is possible for relatively large cracks which was introduced through heavily machining (for depth of cut up to 20 μm/pass). From these results, it was concluded that mullite/ SiC composite could offer improved reliability with out costly polishing by crack-healing treatment after heavily machining. However, the σ_F of the machined specimen healed was found to decrease slightly with increasing cut depth. This is due to the increase in size of surface chipping as shown in figure 2 (b-1) ~ (b-3).

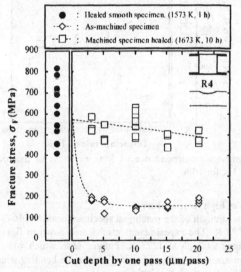

Figure 6. Effect of cut depth by one pass on fracture stress at room temperature of machined specimen healed at 1673 K for 10 h.

Effect of test temperature on the fracture stress of machined specimens healed

Figure 7 shows the temperature dependence of the bending strength of the healed machined specimens. The symbol □ shows the σ_F of the machined specimen healed at 1673 K for 10 h (d = 10 μm/pass). The σ_F of the healed machined specimens decrease almost linearly in the temperature range from room temperature to 1573 K, and these values were found to be more than approximately 350 MPa. Above 1573 K, σ_F decrease slightly with the increase of the test temperature. Moreover, SEM observations confirmed that all of the machined specimens did not fracture from examples of the healed machining-cracks. Thus, it could be concluded that all crack-healed zones exhibited the same levels strength to the base material (mulite / SiC composite) up to 1673 K.

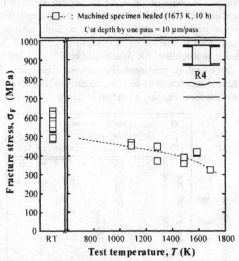

Figure 7. Temperature dependence of fracture stress of machined specimen healed at 1673 K for 10 h.

Static fatigue strength at high temperature

The static fatigue strength of the machined specimen healed at 1673 K for 10 h has also been investigated at 1273 K. The experimental results are shown in figure 8. The symbol □ indicates the static fatigue strength at 1273 K. Fatigue tests which were terminated prior to failure are marked by an arrow (→). The measured monotonic bending strengths (σ_F) under the same temperatures are shown in the left column.

The value of the static fatigue limit (σ_{s0}) at 1273 K for the healed machined specimens was found to be 375 MPa. This value almost equates to the minimum σ_F measured under monotonic bend and is approximately 90 % of the average σ_F of all bend specimens. Moreover, SEM observations confirmed that the machined specimens healed did not fracture from examples of the healed machining-cracks. From this result, it is concluded that all crack-healed zones exhibited the same levels static fatigue strength as the base materials at 1273 K.

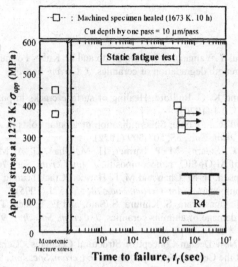

Figure 8. Static fatigue strength at 1273 K of machined specimen healed at 1673K for 10 h.

CONCLUSION

In this study, mullite/ 15 vol.% SiC particles composite was sintered and the crack-healing behavior for the cracks caused by machining was investigated. Small bending specimens (22 × 4 × 3 mm) were made. These specimens were machined into semi-circular groove so that they have stress concentration factor to replicate that of components. The machined specimens were crack-healed under various conditions. The fracture stress and static fatigue strength of these specimens after crack-healing were evaluated systematically. From the obtained results, the crack-healing effect on the cracks was discussed. The main conclusions were as follows:

(1) Fracture stress (σ_F) of machined specimen healed at 1673 K for 10 h was almost equal to that of healed smooth specimens. From these results, the optimized crack-healing conditions for machining cracks was defined as heating at 1673 K for 10 h.
(2) Numerous machining cracks were completely healed by the optimized crack-healing treatment. However chipping introduced at the surface during heavily machining could not be healed, resulting in slightly decreased strength.
(3) Crack-healing was possible for relatively large cracks which was introduced through heavily machining for depth of cut by one pass up to 20 μm/pass.
(4) All crack-healed zone exhibited the same levels strength to the base material (mulite / SiC composite) up to 1673 K.
(5) All crack-healed zones exhibited the same levels static fatigue strength as the base materials at 1273 K.

(6) Therefore, the crack-healing treatment is an effective technique for increasing the reliability of machined mullite/ SiC composite ceramics.

REFERENCES

[1] W. Kanematsu, Y. Yamauchi, T. Ohji, S. Ito and K. Kubo, Formulation for the effect of surface grinding on strength degradation of ceramics, *J. Ceram. Soc. Jpn.*, **100**, 775-779 (1992), (in Japanese).

[2] F. F. Lange and K. C. Radford, Healing of surface cracks in polycrystalline Al_2O_3, *J. Am. Ceram. Soc.*, **53**, 420-1 (1970).

[3] C. F. Yen and R. L. Coble, Spheroidization of tubular voids in Al_2O_3 crystals at high temperatures, *J. Am. Ceram. Soc.*, **55**, 507-509 (1972).

[4] J. Zhao, L. C. Stearns, M. P. harmer, H. M. Chan, G. A. Miller and R. F Cook, Mechanical behavior of Al_2O_3-SiC "nanocomposite, *J. Am. Ceram. Soc.*, **76**, 503-510 (1993).

[5] A. M. Thompson, H. M. Chan and M. P. Harmer, Crack healing and surface relaxation in Al_2O_3-SiC "Nanocomposite, *J. Am. Ceram. Soc.*, **78**, 567-571 (1995).

[6] Y. Matsuo, T. Ogasawara, S. Kimura, S. Sato, and E. Yasuda, The effect of annealing on surface machining damage of alumina ceramics, *J. Ceram. Soc. Jpn.*, **99**, 384-389 (1991), (in Japanese).

[7] Niihara, K., New Design Concept of Structural Ceramics - Ceramic Nanocomposites, The Chemical Issue of the Ceramic Society of Japan, *J. Ceram. Soc. Jpn.*, **99**, 974-82 (1991).

[8] B. S. Kim, K. Ando, M. C. Chu and S. Saito, Crack-healing behavior of monolithic alumina and strength of crack-healed member, *J. Soc. Mater. Sci. Jpn.*, **52**, 667-673 (2002), (in Japanese).

[9] K. Ando, B.S. Kim, M.C. Chu, S. Saito and K. Takahashi, Crack-healing and mechanical behavior of Al_2O_3/SiC composites at elevated temperature, *Fatigu. Fract. Engin. Mater. Struct.*, **27**, 533-541 (2004).

[10] M. C. Chu, S. Sato, Y. Kobayashi and K. Ando, Study on Strengthening of Mullite by Dispersion on Carbide Ceramics Particles, *Jpn. Soc. Mech. Eng.*, **60**, 2829-2834 (1994) (in Japanese)

[11] S. Sato, M. C. Chu, Y. Kobayashi and K. Ando, Strengthening of Mullite by Dispersion of Carbide Ceramics Particles (2nd Report, Effect of SiC grain Size and Heat Treatment), *Jpn. Soc. Mech. Eng.*, **61**, 1023-1030 (1995). (in Japanese)

[12] M. C. Chu, S. Sato, Y. Kobayashi and K.Ando, Damage Healing and Strengthening Behavior in Intelligent Mullite/ SiC Ceramics, *Fatigue Fract. Engng. Mater. Struct.*, **18**, 1019-1029 (1995).

[13] K. Ando, K. Hurusawa, M. C. Chu, T. Hanagata, K. Tuji and S. Sato, Crack-healing behavior under stress of mullite silicon carbide ceramics and the resultant fatigue strength, *J. Am. Ceram. Soc.*, **84**, 2073-2078 (2001).

[14] H. Z. WU, S. G. Roberts and B. Derby, The strength of Al_2O_3/SiC nanocomposite after grinding and annealing, *Acta Mater.*, **46**, 3839-48 (1998).

[15] K. Takahashi, M. Yokouchi, S.K. Lee, K. Ando, Crack-healing behavior of Al_2O_3 toughened by SiC whiskers, *J. Am. Ceram. Soc.*, **86**, 2143-2147 (2003).

[16] S. K. Lee, K. Takahashi, M. Yokouchi, H. Suenaga and K. Ando, High temperature fatigue strength of crack-healed Al_2O_3 toughened by SiC whiskers, *J. Am. Ceram. Soc.*, **87**, 1259-1264 (2004).

[17] K. Ando, M. Yokouchi, S. K. Lee, K. Takahashi, W. Nakao and H. Suenaga, Crack-healing behavior, high temperature strength and fracture toughness of alumina reinforced by SiC whiskers, *J. Soc. Mat. Sic., Jpn.*, **53**, 599-606 (2004), (in Japanese).

[18] S.K. Lee, M. Ono, W. Nakao, K. Takahashi, and K. Ando, Crack-healing behavior of mullite/SiC/Y_2O_3 composites and its application to the structural integrity of machined components, *J. Eur. Ceram. Soc.*, **25**, 3495-3502 (2005)

[19] T. Osada, N. Wataru, K. Takahashi, K. Ando and S. Saito, Strength recovery behavior of machined Al_2O_3 / SiC nano-composite by crack-healing, *J. Eur. Ceram. Soc.*, **27**, 3261-3267 (2007)

[20] T. Osada, N. Wataru, K. Takahashi, K. Ando and S. Saito, Strength recovery behaviore of machined alumina / SiC whisker composite by crack-healing, *J. Ceram. Soc. Jpn.*, **115**, 278-284 (2007), (in Japanese).

[21] Japan Industrial Standard R1601, Testing method for flexural srength of high performance ceramics, Japan Standard Association, Tokyo (1993).

[22] M.Nishida, Stress concentration. Morikita Publishing Co., Ltd., 572-574 (1967), (in Japanese).

[23] Japan Industrial Standard R1632, Testing method for static bending fatigue of fine ceramics, Japan Standard Association, Tokyo (2003).

EFFECTS OF OXIDATION ON THE MECHANICAL PROPERTIES OF PRESSURELESS-SINTERED SiC-AlN-Y₂O₃ COMPOSITES OBTAINED WITHOUT POWDER BED

EFFECTS OF OXIDATION ON THE MECHANICAL PROPERTIES OF PRESSURELESS-SINTERED SiC-AlN-Y$_2$O$_3$ COMPOSITES OBTAINED WITHOUT POWDER BED

G.Magnani
ENEA
Dept. of Physics Tech. and New Materials
Bologna Research Center
Via dei Colli 16
40136 Bologna
Italy

L.Beaulardi, E.Trentini
ENEA
Dept. of Physics Tech. and New Materials
Faenza Research Center
Via Ravegnana 186
48018 Faenza (Ra)
Italy

ABSTRACT

Mechanical properties of SiC-AlN-Y$_2$O$_3$ composites (SiC 50%wt-AlN 50%wt), pressureless-sintered with an innovative and cost-effective method, were determined before and after oxidation performed at 1300°C for 1h. As a consequence of the oxidative treatment, fracture toughness increased from 4.6 MPa m$^{1/2}$ to 6.6 MPa m$^{1/2}$, flexural strength from 420 MPa to 488 MPa, Weibull modulus from 4.5 to 5.3 and thermal shock resistance (expressed as critical temperature difference) from 310°C to 380°C. First of all, these results demonstrated that a pre-oxidation treatment is needed to increase the mechanical resistance and reliability of SiC-AlN-Y$_2$O$_3$ components. Secondarily, the beneficial effects of the oxidation on the mechanical properties could be explained in terms of compressive residual stresses and crack healing ability.

INTRODUCTION

Liquid phase sintered (LPS) silicon carbide is a candidate material for high temperature structural components. Sintering additives such as Al$_2$O$_3$-Y$_2$O$_3$ and AlN-Y$_2$O$_3$ were extensively used to obtain fracture resistant LPS-SiC ceramics[1-4]. In the case of SiC-AlN-Y$_2$O$_3$ composites, previous studies reported that 2H SiC-AlN solid solution with improved high temperature properties could be achieved with SiC/AlN weight ratio greater than 80/20 and above 1850°C[1-2].

In a previous paper, we have already demonstrated that high density SiC-AlN-Y$_2$O$_3$ ceramics (SiC 50%wt-AlN 50%wt) can successfully be obtained by liquid-phase pressureless sintering without using a powder bed[3]. Mechanical properties were also determined and high fracture resistance together to high temperature properties were put in evidence[4]. Like brittle materials, however, strength of this class of materials is closely related to the size and distribution of surface flaws, because of their inherent low toughness.

A well-known method to overcome this problem is based on the crack healing. Studies on this argument have been reported for different types of SiC-based ceramics and the major observation was that crack healing induced by pre-oxidation procedure is able to strengthen and increase the reliability of ceramic components[5-7]. In the above perspectives, the effects of pre-oxidation on mechanical properties and reliability of pressureless sintered SiC-AlN-Y$_2$O$_3$ composites were investigated and reported in the present paper.

EXPERIMENTAL PROCEDURE

Commercially available α-SiC (UF10, H.C. Starck, Germany), AlN (Pyrofine A, Atochem, France), Y$_2$O$_3$ (H.C. Starck, Germany) were used as starting powders. Characteristics of these powders are reported in Table I.

The powder batch was composed by 48%wt SiC, 48%wt AlN and 4%wt Y$_2$O$_3$ and was wet-mixed in ethanol for 12 h using SiC grinding balls. After drying and sieving, the powder was compacted by die pressing at 67 MPa and subsequently was pressed at 150 MPa by cold isostatic press (CIP).

Sintering was performed in a graphite elements furnace in flowing nitrogen at 1 atm with green bodies put inside a graphite crucible without powder bed. Sintering was performed at 1950°C, while an annealing step was conducted at 2050°C. Thermal cycle was characterised by heating and cooling rate of 20-30°C/min and by dwell time of 0.5 h at the sintering temperature.

Mechanical properties were determined before and after oxidation. Fracture toughness was determined by means of Vickers indentation method with indentation load in the range 49-196N[8], whereas thermal shock resistance was determined by water quenching method in accordance with the standard EN 820-3. Finally, flexural strength was determined by four-point bend tests at room temperature. Nine samples as bars of 3 x 4 x 45 mm^3 were prepared and tested in accordance with the standard ENV 843-1 (crosshead speed 0.5 mm/min).

Weibull distribution of strength was also determined by:

$$P_i(\sigma) = 1 - exp\left(-\frac{\sigma^m}{\sigma_0^m}\right) \tag{1}$$

where $P_i(\sigma)$ is the probability of failure at a stress, σ, σ_0 is the characteristic strength and m is the Weibull modulus. By taking the logarithm twice, Equation 1 can be rewritten in a linear form:

$$ln\,ln\left(\frac{1}{1-P}\right) = m\,ln\,\sigma - m\,ln\,\sigma_0 \tag{2}$$

The probability estimator was calculated following method proposed by Tiryakioglu[9] in order to obtain an unbiased estimate of the Weibull modulus for sample sizes between 9 and 50:

$$P_i = \frac{n - 0,13}{N} \tag{3}$$

where N is the total number of specimens tested and n is the specimen rank in ascending order of failure stress.

Table 1. Characteristics of the starting powders

Powder	Purity (wt%)	Specific Surface Area (m^2/g)	Particle size (μm)
α-SiC	>97.0	15.6	0.48
AlN	>97.0	3.6	0.1-0.5
Y$_2$O$_3$	99.9		<5

RESULTS AND DISCUSSION

Fracture toughness

Previous works demonstrated that oxidation products changed on the basis of the SiC-AlN weight ratio and oxidation temperature[10-11]. The main constituents of pressureless-sintered 50%wtSiC-50%AlN oxidized layer were mullite, α-cristobalite and yttrium disilicate (γ-Y$_2$Si$_2$O$_7$)[12]. In addition, it must be taken into account that in this material the oxidation mechanism of SiC changed from passive to active when oxidation temperature is higher than 1300°C[12].

Consequently, oxidation effects on mechanical properties had to be assessed as a function of the oxidation temperature. With this aim, apparent fracture toughness (K_{APP}) was determined after oxidation for 1h in the temperature range 1200-1500°C to evaluate toughening effects induced by the formation of the oxidic species. Results are reported in Figure 1 as a function of the oxidation temperature. K_{APP} of the as-sintered sample was 4.6 MPa m$^{1/2}$, while the same material presented a value of 6.2 MPa m$^{1/2}$ after oxidation at 1200°C for 1h. After oxidation at 1300°C for 1h, the toughening effect, induced by the oxidation, was particularly evident and permitted to reach the best value of 6.6 MPa m$^{1/2}$. At higher temperature the active oxidation of SiC caused the formation of defects with rupture and spallation of the oxide scale[12] due to the formation of the gaseous species SiO and CO. As a consequence, apparent fracture toughness started to decrease above 1300°C and it reached the same value of the as-sintered sample after oxidation at 1400°C. Rixecker et al.[13] reported a similar behavior in the gas-pressure sintered SiC (90%vol)-AlN-Y₂O₃ composites. This material, with a K_{APP} of 4.3 MPa m$^{1/2}$, was treated in air at 1200°C for 12 min and reached a K_{APP} value of 6.1 MPa m$^{1/2}$. As explanation of this increase, the volume gain associated to the oxidation of the intergranular phases (Y₁₀Al₂Si₃O₁₈N₄ before oxidation, YAG and yttrium disilicate after oxidation) was indicated as responsible of the creation of compressive stress on the surface. In particular, this theory was based on the preferential loss of nitrogen from the additive during the heat treatment.

In the present study, the same mechanism should be responsible of the increase of the apparent fracture resistance. SiC-AlN-Y₂O₃ composite was treated at 1300°C for 1h in fluent argon to confirm this theory. In this condition, oxidation of the intergranular phase (Y₁₀Al₂Si₃O₁₈N₄)[4] was completely inhibited and K_{APP} resulted equal to 4.5 MPa m$^{1/2}$, practically the same value of the as-sintered sample.

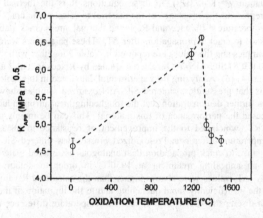

Figure 1. Apparent fracture toughness (K_{app}) as function of the oxidation temperature.

Flexural strength and thermal shock resistance

Figure 2 shows the values of residual strength after thermal shock of as-sintered samples and samples heat-treated at 1300°C for 1h in air. Oxidized samples demonstrated higher flexural strength and higher critical temperature difference, ΔT_c, than as-sintered samples.

The increase of the flexural strength might be caused by the passive oxidation of SiC which enhanced the crack healing mechanism. This phenomenon is typical of SiC-based materials and is based on the production of SiO$_2$ which is able to fill large defects and to be bonded to the base material strongly[14-15]. Furthermore, the bending strength shows a three-stage behavior (Figure 2). In the first stage the thermal shock treatment does not affect this mechanical property. The second stage, corresponds to a significant and sudden decrease of the bending strength. In the third stage the flexure resistance holds a constant level, but significantly lower than that of the first stage. The presence of these three stages is characteristic of the Hasselman's model for the thermal shock[16]. Three thermal stress parameters could be calculated by:

$$R = \sigma (1-v) / E\alpha \tag{4}$$

$$R^{IV} = E\gamma_f \left[\sigma^2 (1-v) \right] \approx \left(K_{IC} / \sigma \right)^2 / (1-v) \tag{5}$$

$$R_{st} = \left[\gamma_f / \left(\alpha^2 E \right) \right]^{\frac{1}{2}} \tag{6}$$

where K_{IC} is the fracture toughness of the material, E is the elastic modulus, α is the thermal expansion coefficient, v is the Poisson's modulus and γ_f is the fracture surface energy (which is determined on the basis of the Irwin's equation $\gamma_f = K_{IC}^2/2E$[17]). In these equations R is the thermal stress fracture resistance parameter, R^{IV} represents the resistance of the material to catastrophic crack propagation of ceramic under a critical temperature difference and R_{st} is the thermal stress crack stability parameter which indicates the resistance to crack repropagation after ΔT_c. These parameters were calculated for as-sintered and oxidized samples and the values are reported in Table 2 together with the data referred to the gas pressure-sintered (GPS) SiC-AlN-Y$_2$O$_3$ material studied by Rixecker et al.[18] and recalculated on the basis of the equations (4)-(6). Analyzing the experimental data listed in this table, the first point which could be drawn it is that pressureless-sintered SiC-AlN oxidized samples showed the highest values of R, R^{IV} and R_{st} as further demonstration that the toughening mechanism induced by the heat treatment at 1300°C increased the performance of this material. This can be mainly assessed on the basis of the increase of R^{IV}, which led to the improvement of resistance to instantaneous crack propagation at critical temperature difference. Pre-oxidized pressureless-sintered SiC-AlN ceramics also exhibited higher resistance to crack propagation than analogous as-sintered material obtained by GPS and toughened by an annealing treatment at 1950°C in order to obtain a platelet-type microstructure. Secondarily, ΔT_c values were always higher than the calculated R values and this can be probably associated to the insufficiently rapid quenching and to the limitation of the water cooling rate which could lead to a discrepancy between the recorded temperature difference and that of the specimen tested[19].

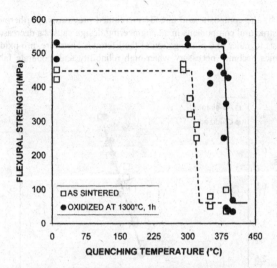

Figure 2. Thermal shock behaviour of as-sintered and oxidized SiC-AlN-Y$_2$O$_3$ ceramics.

Table II. Thermal shock resistance parameters of pressureless sintered SiC-AlN-Y$_2$O$_3$ composites (this study) and GPS SiC-AlN-Y$_2$O$_3$ ceramics (data extracted from Rixecker et al.[18]).

	SiC-AlN-Y$_2$O$_3$ (this work, as-sintered)	SiC-AlN-Y$_2$O$_3$ (this work, after oxidation at 1300°C,1h)	SiC-AlN-Y$_2$O$_3$ (GPS)[18]	SiC-AlN-Y$_2$O$_3$ (GPS, annealed at 1950°C,16h)[18]
E (GPa)	359	359	400	400
ΔT_C (°C)	310	380	455	500
α (10^{-6}/°C)	4.7	4.7	4.3	4.3
ν	0.25	0.25	0.28	0.28
σ (MPa)	420	488	506	466
K_{Ic} (MPa m$^{1/2}$)	4.6	6.6	4.9	6.2
R (°C)	187	217	212	196
R^{IV} (μm)	174	265	130	246
R_{st} (μm$^{1/2}$ °C)	1928	2766	2014	2549

Weibull modulus

The effect of the heat treatment at 1300°C for 1h on reliability of pressureless sintered SiC-AlN ceramics was estimated by means of the determination of the Weibull parameters reported in equation (2). They were determined from the linear adjustment by the least square method of the straight lines reported in Figure 3. Results in term of Weibull modulus (m), characteristic strength values (σ_0) for probability of failure 63% and median strength (σ_{50}) obtained with as-sintered and oxidized samples are summarized and compared in Table III. Even for the small number of samples used, the obtained fits are sufficient to establish significant differences between the materials. Oxidized samples showed the highest values of Weibull modulus, σ_{50} and σ_0. Because Weibull modulus reflects the reliability of brittle materials and homogeneity of the testing data, oxidized SiC-AlN ceramics are more reliable, repeatable and homogeneous than as-sintered samples. This is a direct effect of the crack healing mechanism showed by the composite and already mentioned previously. Furthermore, crack healing behavior applied to structural components in an engineering design causes a decrease in the machining and maintenance costs together to a prolongation of the lifetime. Therefore, pre-oxidation treatment of LPS-SiC-AlN ceramics becomes necessary when high reliability and moderate fabrication costs are required.

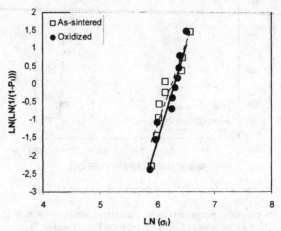

Figure 3. Weibull analysis of the strength values.

Table III. Strength and Weibull modulus of as-sintered and oxidized pressureless-sintered SiC-AlN ceramics.

Material	Weibull modulus (m)	Characteristic strength, σ_0 (MPa)	Median strength, σ_{50} (MPa)
As-sintered	4.5	525	484
Oxidized	5.3	547	510

CONCLUSION

Oxidatiaon treatment at 1300°C led to a considerable increase of the apparent fracture toughness. This phenomenon could be explained on the basis of the surface compressive stresses induced by the volume gain associated to the oxidation of the intergranular phase. Furthermore, the oxidation-induced toughening effect led to the improvement of resistance to crack propagation after a critical temperature difference. Flexural strength and reliability are also positively influenced by the heat treatment at 1300°C since that the pre-oxidized SiC-AlN material showed MOR and Weibull modulus values significantly higher than the as-sintered material.

REFERENCES

[1] A. Zangvil, and R. Ruh, Phase Relationship in the Silicon Carbide-Aluminum Nitride System, *J.Am.Ceram.Soc.*, **71**(10), 884-90 (1988).

[2] C.S. Lim, Effect of a-SiC on the Microstructure and Toughening of Hot-Pressed SiC-AlN Solid Solution, *J.Mat.Sci.*, **35**, 3029-35 (2000).

[3] G. Magnani, and L. Beaulardi, Properties of Liquid Pphase Pressureless Sintered SiC – Based Materials Obtained without Powder Bed, *J.Aus.Ceram.Soc.*, **41**(1), 31-36 (2005).

[4] G. Magnani, and L. Beaulardi, Mechanical Properties of Pressureless Sintered SiC-AlN Composites Obtained Without Sintering Bed, *Ceram.Eng.Sci.Proc.*, **25**(4) 31-36 (2004).

[5] M.C. Chu, S.J. Cho, H.M. Park, K.J. Yoon, and H. Ryu, Crack-healing in reaction-bonded silicon carbide, *Mater.Lett.*, **58**(7-8), 1313- 16 (2004).

[6] M.C. Chu, S.J. Cho, H.M. Park, K.J. Yoon, and H. Ryu, Crack-healing in silicon carbide, *J.Am.Ceram.Soc.*, **87**(3), 490-92 (2004).

[7] M.C. Chu, S.J. Cho, D. Sarkar, B. Basu, G.J. Yoon, and H.M. Park, Oxidation-Induced Strengthening in Ground Silicon Carbide, *J.Mater.Sci.*, **41**, 4978-80 (2006).

[8] K. Niihara, A. Nakahira, and T. Hirai, The Effect of Stoichiometry on Mechanical Properties of Boron Carbide", *J.Am.Ceram.Soc.*, **67**, C-13-C-15 (1984).

[9] M. Tiryakioglu, "An Unbiased Probability Estimator to Determine Weibull Modulus by the Linear Regression Mmethod", *J.Mater.Sci.*, **41**, 5011-13 (2006).

[10] V.A. Lavrenko, M. Desmanion-Brut, A.D. Panasyuk, and J. Desmanion, Features of Corrosion Resistance of AlN-SiC Ceramics in Air Up to 1600°C, *J.Eur.Ceram.Soc.*, **18**, 2339-43 (1998).

[11] D. Sciti, F. Winterhalter, and A. Bellosi, Oxidation behaviour of pressureless sintered AlN-SiC composite, *J.Mater.Sci.*, **39**, 6965-73 (2004).

[12] G. Magnani, and L. Beaulardi, Long Term Oxidation Behaviour of Liquid Phase Pressureless Sintered SiC-AlN Ceramics Obtained Without Powder Bed, *J.Eur.Ceram.Soc.*, **26**(15), 3407-13 (2006).

[13] G. Rixecker, I. Wiedmann, A. Rosinus, and F. Aldinger, High-Temperature Effects in the Fracture Mechanical Behaviour of Silicon Carbide Liquid-Phase Sintered with AlN-Y$_2$O$_3$ Addittives, *J.Eur.Ceram.Soc.*, **21**, 1013-19 (2001).

[14] F.F. Lange, Healing of Surface Cracks in SiC by Oxidation, *J.Am.Ceram.Soc.*, **53**, 290 (1970).

[15] Y. Kim, K. Ando, and M.C. Chu, Crack-Healing Behavior of Liquid-Phase-Sintered Silicon Carbide Ceramics, *J.Am.Ceram.Soc.*, **86** (3), 465-70 (2002).

[16] D.P.H. Hasselmann, Elastic Energy at Fracture and Surface Energy as Design Criteria for Thermal Shock, *J.Am.Ceram.Soc.*, **46**, 535-38 (1963).

[17] G. R Irwin, "Fracture"; p.551 in *Handbuch der Physik*, Vol. 6, Springer-Verlag, Berlin, 1958.

[18] G. Rixecker, K. Biswas, A. Rosinus, S. Sharma, I. Wiedmann, and F. Aldinger, Fracture Properties of SiC Ceramics with Oxynitride Additives, *J.Eur.Ceram.Soc.*, **22**, 2669-75 (2002).

[19] D.C. Jia, Y. Zhou, and T.C. Lei, Thermal Shock Resistance of SiC Whiskers Reinforced Si$_3$N$_4$ Ceramic Composites, *Ceram. Int.*, **22**, 107-12 (1996).

FIBER PUSH OUT TESTING BEFORE AND AFTER EXPOSURE: RESULTS FOR AN MI SIC/SIC COMPOSITE

G. Ojard[2], L. Riester[3], R. Trejo[3], R. Annis[4], Y. Gowayed[5], G. Morscher[6], K. An[3], R. Miller[2], and R. John[1].

[1] Air Force Research Laboratory, Wright-Patterson AFB, OH
[2] Pratt & Whitney, East Hartford, CT
[3] Oak Ridge National Labs, Oak Ridge, TN
[4] APremaTech Chand, Worcester, MA
[5] Auburn University, Auburn, AL
[6] Ohio Aerospace Institute, Cleveland, OH

ABSTRACT

The increased interest in ceramic matrix composites requires the knowledge of how the material behaves after periods of exposure to temperature and stress. An important parameter under consideration is the fiber/coat/matrix interfacial sliding, which has a direct impact on the residual properties of the material after exposure. A series of fiber push-out tests were conducted on Melt Infiltrated SiC/SiC material considered as a strong candidate for hot section applications in gas turbine engine especially in the combustor and turbine. Samples were exposed under conditions of creep or dwell fatigue at 1204°C for periods ranging from 10 hours to 2,000 hours under stresses of 110.4 MPa or 165.6 MPa. As part of this effort, the sample configuration was changed from the standard wedge shape sample to a uniform cross section where the sample was taken out of standard microstructural sample mounts. This allowed more samples to be manufactured in a more consistent manner. The results of the sample change from a transmission electron microscopy sample to a uniform cross section sample as well as the effects of exposure will be presented and discussed.

INTRODUCTION

There is ever increasing interest in Ceramic Matrix Composites (CMCs) due to the fact that mechanical properties are relatively constant with temperatures up to the maximum use temperature [1]. This is shown in the interest of CMCs for extended high temperature use where superalloys are usually considered [2,3]. As characterization of this class of material proceeds to enable such applications, key constituent properties need to be looked into in both the as-received state as well as after relevant exposure. The purpose of this paper is look at the interfacial shear stress of a known CMC system in both cases just mentioned and to additionally recommend a new sample design to make this property easier to measure.

The fiber/matrix interface is a key area of study since the interface controls if the CMC behaves as a composite or not [4]. To assure composite behavior a weak bond is desired and this is the reason Carbon or Boron Nitride is used as the fiber interface coating [5]. The interfacial shear stress is a key property since it controls (influences) the prevalent damage mechanism and the resulting non-linearity [6-7].

While there are several tests that can be done to determine the interfacial shear stress [4], one of the better known is fiber push in testing [4,6]. It is known from the work of others [8], that the sample preparation for this is very difficult. Part of the work for this paper is to show a different method for sample fabrication that allows a larger and more uniform sample to be made for this test.

In addition, not only is the interfacial shear stress important to know in the as-received state, it as well as other parameters need to be known as the material is used in service. More importantly, this needs to be known under durability conditions (testing) of stress and temperature. For the MI SiC/SiC system for this paper, all of the testing with exposure was done at 1204°C and stresses around 165.6 MPa. This temperature is near the maximum use temperature of the material [9] and the stress is within the proportional limit of the material [10].

PROCEDURE

Material Description
For this testing, the composite system interrogated was a Melt Infiltrated In-Situ BN SiC/SiC composite (MI SiC/SiC). The interface coating for this material is form a two step process: it is initially heat treated to create a fine layer in in-situ BN and then it is followed with CVI deposited Si-doped BN. The MI SiC/SiC system has a stochiometric SiC (Sylramic™) fiber in a multiphase matrix of SiC deposited by chemical vapor deposition followed by slurry casting of SiC particulates with a final melt infiltration of Si metal. The specific MI SiC/SiC tested for this effort had 36% volume fraction fibers using a 5 HS weave at 20 EPI. The fibers are 10 μm diameter and there are 800 fibers per tow. This material system was developed by NASA-GRC and is sometimes referred to as the 01/01 material [11]. A cross section of this material is shown in Figure 1.

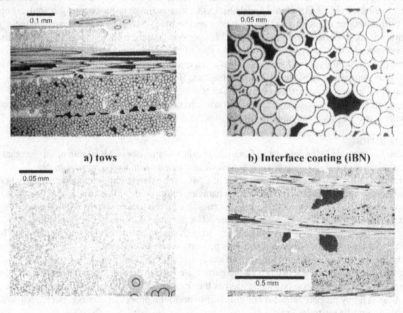

a) tows   b) Interface coating (iBN)

c) SiC particulate with Si   d) porosity
Figure 1. MI SiC/SiC Microstructure Images

Sample Fabrication

The sample fabrication process starts with a sample of interest mounted within epoxy (or other mount material) as a standard microstructural sample for optical and SEM cross-section work. The typical mount is approximately 38 mm in diameter and around 20 mm in thickness. A series of steps were undertaken to arrive at the final fiber push out sample and the equipment for this effort can be found in a typical machine shop knowledgeable in the machining and preparation of ceramic test specimens or components. During the sample fabrication for this effort as well as past microscopy samples, there was no indication of epoxy infiltrating into the material that could influence any testing results. Considering the low porosity of this material, it is not surprising that this was not seen. (This may be an issue for other CMC systems.)

The sample process starts with a slice cut from the mount approximately 1.0 mm thick. This was done with a standard surface grinder equipped with a standard cut off wheel. A synthetic water based coolant was used. The sample was mounted to a steel pin with double-sided tape and then the OD was machined down to approximately 16 mm keeping the sample of interested centered. It is not crucial what method is used to reduce the epoxy mount diameter. The OD is machined to aid in later polishing steps. The sample is then reduced in thickness to 0.7 mm with a standard surface grinder with a 320 grit resin bonded wheel (with coolant). This ensures that the specimen faces are parallel. (This method produces a constant cross section sample.) The specimens are then mounted to a steel fixture plate using a low temperature bonding adhesive. (For this effort, three samples were mounted and were evenly spaced on the steel plate that was 76 mm in diameter.) The thicknesses of the specimens were reduced to 0.5mm with a standard surface grinder equipped with a 320 grit resin bonded wheel using coolant. Reliefs were then machined 0.13mm deep on one face of each specimen, see Figure 2. The reliefs reduce the surface area that will require polishing, thus making the polishing steps easier. (Dimensions of the relief are not critical.) The specimens were polished by hand using diamond compound. Polishing started with 9 micron compound, progressively getting finer using 6, 3, and finally 1 micron. Total material removal in the polishing process was approximately 0.05mm. The final specimen thickness target was between 0.5 and 0.3mm. The thickness of the samples for this effort was an initial thickness of 0.7 mm but most work was on samples that were 0.35 mm. The specimens were removed from the steel plate using a low temperature heat source and then cleaned. Due to the relatively thin cross section, the CMC piece may "fall out" of the mount material and there may be warping of the epoxy mount material. The sample can be tested if it "falls out" and the warping was not an issue in the test as it was slight.

Figure 2. Schematic of sample showing reliefs

Fiber Push Out

Fiber push out was done using a Nano-Indentor II. The nano-indentor had a 6 μm flat bottom punch for the tip and load versus displacement was recorded during the test electronically. The maximum load during testing was 450 mN. The data was reviewed by plotting the Force2 versus displacement [5]. A schematic for the test results are shown in Figure 3. Where the fiber slips is where the interfacial shear stress is determined. This is determined by the following equation:

$$\tau = F/2\pi rt$$

where τ is the interfacial shear stress, F is the force when the fiber slips, r is the radius of the fiber and t is the sample thickness.

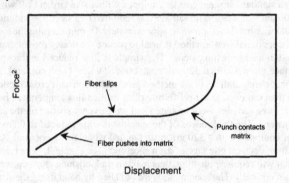

Figure 3. Schematic of Fiber Push Out Test [5]

Samples for Testing

Several samples were fabricated from as-received material to not only determine the as-received interfacial shear stress but also as an aid in practicing the sample fabrication process. After initial trials of both machining and fiber push out testing, it was decided to settle on a sample thickness of 0.35 mm. In addition to the as-received material, several samples were made available that had been tested at 1204°C with a hold stress of 165.6 MPa in air for periods of 4, 250 and 1508 hours. (The test was either a creep or a 2 hour dwell fatigue test [12].) All samples were machined per the procedure outline above. All samples for this effort were the same thickness to eliminate any issue about residual stresses being a factor due to machining.

RESULTS

Sample fabrication

Sample fabrication produced very uniform samples with a high quality surface finish. This is shown in Figure 4. This figure clearly shows that the sample started as an optical/SEM mounted sample. In addition, the resulting surface finish used for this procedure is of typical optical quality.

 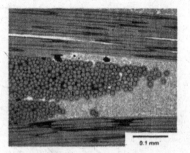

a) Macro image of sample   b) Micro image of sample finish

Figure 4. Typical images of samples used for fiber push out testing

Fiber Push Out – As-Received

Several attempts were made to push out the fibers using the nano-indentor (15 attempts) and there were only three successful attempts were achieved where the data could be interpreted. These successful fiber push out curves are shown in Figure 5. The push out attempts that were not successful were due to the punch hitting the CVI SiC surrounding the fiber indicating that alignment of the punch onto the center of the fiber is crucial. For this data set, the average interfacial shear stress was determined (via equation above) to be 30.1 MPa (6.1 MPa Standard Deviation). This range is within the values published by other authors [7].

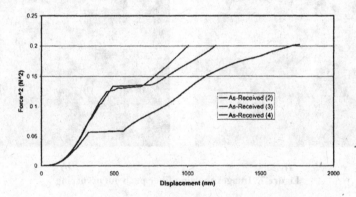

Figure 5. Fiber push out curves performed in the as-received state

For this effort, only the exposed sample that was extracted at the failure face of a creep sample that was tested at 165.6 MPa at 1204°C for 1508 hours was tested. There were 5 successful fiber push outs out of 13 attempts. These successful fiber push out curves are shown in Figure 5. For this data set, the average interfacial shear stress was determined (via equation above) to be 17.8 MPa (3.8 MPa Standard Deviation). Again, alignment onto the fiber is key.

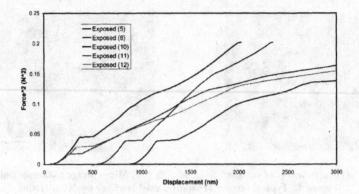

Figure 6. Fiber push out curves performed in the exposed state
(1508 hrs, 165.6 MPa, 1204°C, sample failed)

Some micrographic examples of fiber push out testing on the exposed sample is shown in Figure 7. Figure 7 clearly shoes that the fiber push out is occurring. It is also clear in the images that the punch (indentor) is hitting the CVI SiC matrix around the fiber after push out is occurring giving rise to the increase in load consistent with Figure 3 [5] after the fiber slide.

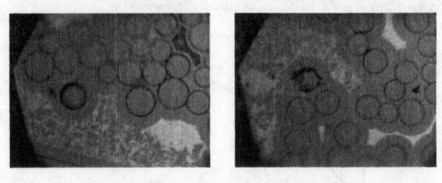

a)   b)
Figure 7. Images showing fiber push out occurring

The full data results for this effort are shown in Table I. A review of the data using statistical tools shows that the data is two statistically different populations. The interfacial shear stress is higher in the as-received state than the exposed state.

Table I. Push Out Fiber Results wit calculated Interfacial Shear Stress

Condition	Test Sequence (#)	Force (N)	Interfacial Shear Stress (MPa)
As-Received	2	0.365	32.2
As-Received	3	0.353	34.9
As-Received	4	0.237	23.2
As-Received	**Average**		**30.1**
As-Received	StDev		6.1
Exposed	5	0.197	19.9
Exposed	8	0.197	21.6
Exposed	10	0.212	18.3
Exposed	11	0.131	11.5
Exposed	12	0.170	17.5
Exposed	**Average**		**17.8**
Exposed	StDev		3.8

DISCUSSION

The as-received interfacial shear stress was determined to be 30.1 MPa. The interfacial shear stress has been determined by other authors for the Sylramic Fiber with BN interface coating and the value was found to be 10 to 70 MPa depending on the type of bonding resulting for the BN and insitu-BN as seen in tensile tests [7]. The value shown here is consistent with the "outside debonding" interfacial shear strength reported by others [7] even though this particular set of samples has shown "inside debonding" as shown in Figure 8 [13]. The iBN pulls away from the fiber upon failure as shown in Figure 8.

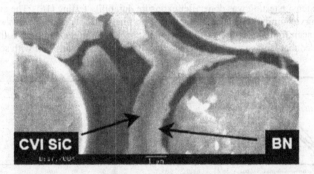

Figure 8. Fracture face from tensile test showing inside debonding

The exposed sample shows a decrease in the interfacial shear strength. The interfacial shear strength drops from an average of 30.1 MPa to 17.8 MPa. This indicates that the interface is being environmentally attacked during the durability test due to a combination of stress and environment. This could be due to the BN being oxidized during testing to a species that allows a lower interfacial stress such as B_2O_3 or due to wear at the interface [14,15]. This later approach can be stated as progressive wear at the interfaces between the fiber/coat and the coat/matrix is considered responsible for depreciation of the ability to carry interfacial stresses that starts in limited zones and progresses to cover most of the composite. The approach, as defined by Reynaud [14,15], can be illustrated by two scenarios as shown in Figure 9:

1. Increase in the area of hysteresis loops with the number of cycles leading to a decrease in ability to carry interfacial shear stress (τ) resulting from a local sliding at the fiber/coat or coat/matrix interface.
2. Decrease in the area of hysteresis loops with number of cycles leading to a decrease in the ability to carry interfacial shear stress resulting from a global sliding at all the interfaces.

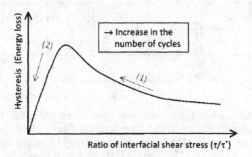

Figure 9. Decrease in interfacial shear stress with mechanical hysteresis following scenario 1 (right) and scenario 2 (left) [14]

Cyclic tensile testing was done after exposure durability testing [16]. The as-received and an exposed curve (similar time and stress of exposed fiber push out sample) are shown in Figure 10. As can be seen from the cyclic tensile tests performed at room temperature: both as received and post exposed that the exposed sample has a smaller hysteresis consistent with scenario 2 above.

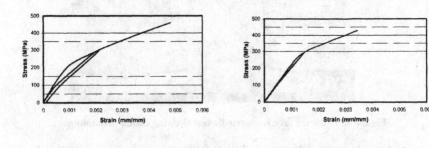

a) As-received tensile test

b) Post exposure tensile test
(165.6 MPa, 1204°C, 250 hours)

Figure 10. Room temperature cyclic tensile tests

CONCLUSION

The as-received interfacial shear strength was determined for the MI SiC/SiC system and was found to be 30.1 MPa. This was done by using a new sample design that allowed a sample to be taken from an optical/SEM mounted sample. This would allow both an optical interrogation of the sample prior to any push out testing. (This was not done for this effort due to time limitations.) The new sample design allows samples to be made with more control on the final sample and can be done using a facility with standard ceramic machining capability. By using a combination of a Nano-indentor and the new sample design, there is a path to generate more data due to the control given from the sample design (uniform thickness and increased sample size) along with the automation that the Nano-indentor provides.

In addition, the interfacial shear strength was determined after exposure and was determined to be 17.8 MPa. This was done for one sample that had been creep tested in air for 1508 hours at a hold stress of 165.6 MPa. This decrease in interfacial shear strength is consistent with the post exposed sample testing that has been done and theory [14,15].

Future work needs to be done on more samples with exposure to determine the rate that the interfacial shear strength degrades. In addition, testing needs to be done to confirm the as-received value determined here against work of others to resolve if there are any issues between materials reported or if it is a material lot effect. This effort should also be expanded to additional material systems to confirm that the sample design can be used more widely. In addition, to make the data more definitive, the other side of the sample should be reviewed to prove conclusively that the fiber was not only pushed in but that is was also pushed out.

ACKNOWLEDGMENTS

The Materials & Manufacturing Directorate, Air Force Research Laboratory under contract F33615-01-C-5234 and contract F33615-03-D-2354-D04 sponsored portions of this work. Research work at ORNL was sponsored by the Assistant Secretary for Energy Efficiency and Renewable Energy, Office of FreedomCAR and Vehicle Technology Program, as part of the High Temperature Materials Laboratory User Program, Oak Ridge National Laboratory managed by UT-Battelle, LLC for the U.S. Department of Energy under contract number DE-AC05-00OR22725

REFERENCES

[1] Wedell, James K. and Ahluwalia, K.S., "Development of CVI SiC/SiC CFCCs for Industrial Applications" 39th International SAMPE Symposium April 11- 14, 1994, Anaheim California Volume 2 pg. 2326.

[2] Brewer, D., Ojard, G. and Gibler, M., "Ceramic Matrix Composite Combustor Liner Rig Test", ASME Turbo Expo 2000, Munich, Germany, May 8-11, 2000, ASME Paper 2000-GT-0670.

[3] Calomino, A., and Verrilli, M., "Ceramic Matrix Composite Vane Sub-element Fabrication", ASME Turbo Expo 2004, Vienna, Austria, June 14-17, 2004, ASME Paper 2004-53974.

[4] Chawla, Chapter 5, Page 169

[5] Morscher, G N; Yun, H M; DiCarlo, J A; and Ogbuji, L T , "Effect of a Boron Nitride Interphase that Debonds between the Interphase and the Matrix in SiC/SiC Composites" J.Am.Ceram.Soc. Vol. 87, no. 1, pp. 104-112. 2004

[6]Evans, A.G., Zok, F.W. and Mackin, T.J., "The Structural Performance of Ceramic Matrix Composites", in **High Temperature Mechanical Behavior of Ceramic Composites**, Nair, S.V. and Jakus, K. Editors, Butterworth-Heinemann, Newton, MA, 1995

[7]Morscher, G.N. and Eldridge, J.I., "Constituent Effects on the Stress-Strain Behavior of Woven Melt-Infiltrated SiC Composites", NASA Technical Report No. 20020024450.

[8]Cranmer, D.C., "Critical Issues in Elevated Temperature Testing of Ceramic Matrix Composites", in **High Temperature Mechanical Behavior of Ceramic Composites**, Nair, S.V. and Jakus, K. Editors, Butterworth-Heinemann, Newton, MA, 1995)

[9]Calomino, A., "Ceramic Matrix Composite (CMC) Materials Characterization" NASA Technical Report No. 20050214034

[10]Y. Gowayed, G. Ojard, J. Chen, R. Miller, U. Santhosh, J. Ahmad and R. John. "Time-Dependent Response of MI SiC/SiC Composites Part 2: Samples with Holes", published in the proceedings of Ceramic Engineering and Science Proceedings, 2007.

[11]Hurwitz, F.I., Calomino, A.M., McCue, T.R., and Morscher, G.N., "C-Coupon Studies of SiC/SiC Composites Part II: Microstructrual Characterization", Ceramic Engineering and Science Proceedings, Vol. 23, Issue 3, 2002, pg 387.

[12] Ojard, G., Gowayed, Y., Chen, J., Santhosh, U., Ahmad J., Miller, R., and John, R., "Time-Dependent Response of MI SiC/SiC Composites Part 1: Standard Samples" published in the proceedings of Ceramic Engineering and Science Proceedings, 2007.

[13]Ojard, G., Miller, R., Santhosh, U., Ahmad, J., Gowayed, Y. and John, R., "MI SiC/SiC Part 1 – Testing Results", 29th Annual Conference on Composites, Materials and Structures, Cape Canaveral/Cocoa Beach, FL. 2005.

[14]Reynaud, P., Rouby, D. and Fantozzi, G., "Effects of temperature and of oxidation on the interfacial shear stress between fibres and matrix in ceramic-matrix composites", Acta mater., vol 46, No. 7, pp 2461-2469, 1998

[15]Reynaud, P., "Cyclic fatigue of ceramic-matrix composites at ambient and elevated temperatures", Composites Science and Technology, vol. 56, pp. 809-814, 1996.

[16]Ojard, G., Calomino, A., Morscher, G., Gowayed, Y., Santhosh, U., Ahmad J., Miller, R. and John, R., "Post Creep/Dwell Fatigue Testing of MI SiC/SiC Composites", published in the proceedings of Ceramic Engineering and Science Proceedings, 2007.

NEW CERAMICS SURFACE REINFORCING TREATMENT USING A COMBINATION OF CRACK-HEALING AND ELECTRON BEAM IRRADIATION

Wataru Nakao, Youhei Chiba, Kotoji Ando
Yokohama National University,
79-5, Tokiwadai, Hodogaya-ku,
Yokohama, 240-8501, Japan
Keisuke Iwata, Yoshitake Nishi
Tokai University
1117, Kitakaname,
Hiratsuka, 259-1292, Japan

ABSTRACT

New surface reinforcement using a combination of crack-healing and electron beam irradiation is proposed for structural ceramics. The influence of electron beam irradiation on the surface hardness was investigated in the present study.

Hot-pressed alumina-30 vol. % SiC particles composite was crack-healed, via a high-temperature oxidation treatment in air at 1573 K for 1 h. Then, the specimens passed through an electron beam curtain to homogeneously irradiate the crack-healed specimens. The total dose of irradiation was controlled by repeating the procedure. Depth profiles of hardness were measured using Vickers indentation with several indented loads.

The hardness of the electron beam irradiated specimens increased from that of crack-healed specimens without electron beam irradiation in the vicinity of the surface. The depth influenced by electron beam irradiation was found to be about 2 μm. Irradiation of 0.432 MGy caused a 38 % hardness increment to surface of the crack-healed specimens. A similar irradiation effect was achieved in non crack-healed specimen. The hardness in the vicinity of the surface varied with electron beam irradiation dose and the optimized conditions of electron beam irradiation to crack-healed and non crack-healed specimens were determined to be 0.432 MGy and 0.216 MGy, respectively.

INTRODUCTION

Surface reinforcements are applied to various mechanical components. One of the typical examples is shot peening for the automotive components made of steel. Giving the surface layer of the treated components large compressive residual stresses, shot peening[1] improves fatigue strength of the components significantly. Therefore, surface reinforcements play important roles to the metallic mechanical components. However, no useful surface treatments have been developed for the structural ceramics yet.

For the surface reinforcements of structural ceramics, it is necessary to control the existence of surface cracks, because fracture strengths of ceramics are well known to be too sensitive to surface cracks. Moreover, managing the existence of surface cracks can be divided between the elimination of cracks introduced before service and the suppression of surface cracking during service.

One of the effective ways to eliminate surface cracks is crack-healing. Especially, that driven by oxidation of SiC can completely erase all surface cracks by a simple heat treatment. Ando et al.[2] reported that a heat treatment at 1573 K for 1 h in air can completely heal a elliptical surface crack with surface length = 100 μm, which introduced by indentation in alumina reinforced by 15 vol.% SiC particles. Also, Osada et al.[3] reported that the numerous surface cracks introduced by heavily machining can be completely healed by the same crack-healing treatment.

On the other hand, low energy electron beam irradiation is expected to suppress surface cracking. Nishi and co-workers[4-6] reported that the electron beam irradiation improved hardness and brittleness of both carbon fiber, CFRP (carbon fiber reinforced polymer) and soda glass, because it causes generation or disappearance of the dangling bonds and vacancy. Moreover, the treatment was reported to have a clearing effect for misted dental mirror glass[7, 8] was reported. The resistance of surface cracking must be directly related to ideal strength. From Orowan's equation, one can understand that the ideal strength of materials is one tenth of elastic modulus. Indeed, hardness has proportional relation to elastic modulus. Therefore, the hardening effect by low energy electron beam irradiation must enhance the resistance to surface craking.

In the present study, a hybrid surface reinforcement consisting of crack-healing and low energy electron beam irradiation has been proposed. The usefulness of the method was investigated from the surface hardness and the fracture strength.

EXPERIMENTAL
Sample Preparation and hybrid surface reinforcement

In the present study, a hot-pressed alumina- 30 vol.% SiC particles composite, which is abbreviated as AS30P in this paper, was used. The hot-pressed AS30P was cut into rectangular test specimens (Width = 4 mm, Length = 22 mm, Height = 3 mm). The detail of the hot pressing is reported elsewhere[9].

The rectangular test specimens were subjected to a hybrid surface reinforcement consisting of low energy electron beam irradiation and crack-healing treatment. The test specimens were crack-healed at 1573 K for 1 h in air. The heat treatment completely eliminates a semi-elliptical pre-crack (surface length = 100 μm, aspect ratio = 0.9) introduced in the center of the specimen surface. For low energy electron beam irradiation, an electro beam curtain system (Type CB175/15/180L, Iwasaki Electric Co. Ltd, Tokyo, Japan) was used. The accelerated voltage and the irradiation current were 170 kV and 2.0 mA, respectively. The irradiation atmosphere was nitrogen with an oxygen content of less than 400 ppm under atmospheric pressure to separate the irradiation chamber from the electron beam generator, which was set under a vacuum atmosphere. Passing thorough the generated electron curtain at a rate of 0.16 m/s, the test specimens were subjected to homogeneous irradiation. The total irradiation dose, D, was varied from 0.216 MGy to 1.73 MGy by repeating the irradiation. The treated specimen is called a hybrid surface reinforced specimen in the paper.

Test Method

The hardness of the hybrid surface reinforced layer was investigated as a function of Vickers indented force, P_V. The value of P_V was varied from 0.490 N to 19.6 N. The diagonal length of Vickers indentation was measured by SEM observation. 3 or 4 indentations were introduced for each P_V.

The fracture strength was investigated using a three point bending with a span of 16 mm. The maximum tensile stress was applied to the pre-crack in the hybrid surface treated layer.

In comparison, the same investigations were conducted on the specimens that are subjected to the crack-healing alone and the low energy electron beam irradiation alone. These specimens were called crack-healed specimens and EB irradiated specimen, respectively. Moreover, the same investigations were also conducted on the smooth specimen, which has not undergone any surface treatment, i.e. it is an as-polished specimen.

RESULTS AND DISCUSSIONS
Surface Hardening Behavior by Low Electron Beam Irradiation

Figure 1 shows the Vickers hardness, HV, of the hybrid surface reinforced specimen ($D = 0.432$ MGy) as a function of the penetration depth of the indenter, D_V, thereby giving a hardness depth profile: The closed and open triangles indicate the measured Vickers hardness for the hybrid surface reinforced specimens ($D = 0.432$ MGy) and crack-healed specimens, respectively. The points demonstrate the data for each measurement. Moreover, it was found that the measured hardness of the crack-healed specimens is independent of the penetration depth, and the indented force, P_V.

As shown in Fig. 1, low energy electron beam irradiation hardened the treated surface layer to a indentation penetration depth of about 3 μm. The electron-beam hardening decayed significantly with increasing penetration depth. Therefore, the hybrid surface reinforced specimen has a large gradient of hardness along the depth direction. However, the enhanced

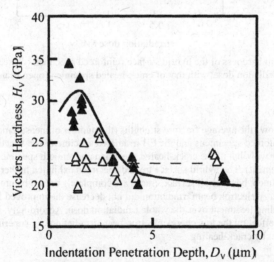

Figure 1 Vickers hardness as a function of penetration depth: hybrid surface reinforced specimens, with $D = 0.432$MGy (closed triangle) and crack-healed specimen (open triangle)

hardness has a maximum of 30.3 GPa at $D_V = 1.6$ μm, corresponding to $P_V = 0.980$ N. The maximum hardness corresponded to a value 1.38 times the value of the specimens without the low energy electron beam irradiation.

Figure 2 shows the relation between the irradiation dose, D, and the maximum hardness of the hybrid surface reinforced specimens. The maximum hardness increases with the irradiation dose increasing below 0.432 MGy, above which, it decreases slightly. From the results shown in Figs. 1 and 2, low energy electron beam irradiation was found to exhibit the no additional effect above a total irradiation dose of 0.432 MGy.

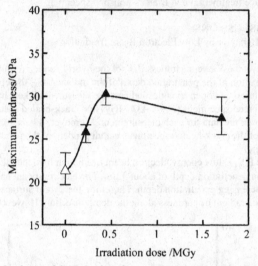

Figure 2 Maximum hardness of the hybrid surface reinforced specimen (closed triangle) as a function of the irradiation dose, with that of crack-healed specimen (open triangle).

Fracture Strength

Figure 3 shows the average fracture strengths (the number of measurement = 3) of the hybrid surface reinforced specimens and the EB irradiated specimens as a function of the irradiation dose, D, with that of the crack-healed specimen and smooth specimen (n is the number of measurement). The hybrid surface treated specimen exhibited higher strength than the EB irradiated specimen, because all surface cracks are completely erased by crack-healing. Moreover, low energy electron beam irradiation did not decrease the improved fracture strength from the crack-healing treatment over the whole irradiation dose. Accordingly, it is concluded that the hardening effect by the low energy electron beam irradiation can superimpose on the strengthening effect by crack-healing.

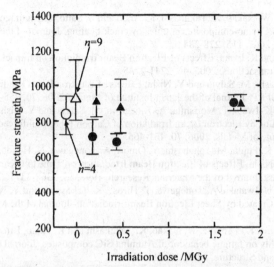

Figure 3 Fracture strength of the hybrid surface reinforced specimen (closed triangle: n=3) as a function of the irradiation dose, with that of EB irradiated specimen (closed circle: n=3). crack-healed specimen (open triangle: n= 9) and smooth specimen (open circle: n=4), where n is the number of measurement.

CONCLUSIONS

A new hybrid surface reinforcement consisting of crack-healing and low energy electron beam irradiation is proposed for structural ceramics. The usefulness of this treatment was investigated in the present study. From the investigation, the following results were derived.

(1)   The hybrid surface reinforcement enhanced the hardness of the treated surface layer. The enhanced hardness exhibited large gradient along the depth direction.

(2)   The maximum hardness of the hybrid surface treated specimen was 1.38 times hardness of the specimen crack-healed alone.

(3)   The hardening effect by the low energy electron beam irradiation can superimpose on the strengthening effect by crack-healing, because the hybrid surface treated specimen has constant improved strength over the whole irradiation dose.

REFERENCES

[1]A. Tange and K. Ando, Improvement of Spring Fatigue Strength by New Warm Stress Double Shot Peening Process. Material Science and Technology. 2002, 18, 642-648
[2] K. Ando, B.S. Kim, M.C. Chu, S. Saito and K. Takahashi, Crack-healing and Mechanical Behaviour of Al_2O_3/SiC Composites at Elevated Temperature. Fatigue and fracture of Engineering Materials and structures. 2004, 27, 533-541

[3]T. Osada, W. Nakao, K. Takahashi, K. Ando and S. Saito, Strength recovery behavior of machined Al$_2$O$_3$/SiC nano-composite ceramics by crack-healing. Journal of the European Ceramics Society. 2007, 115, 278-284

[4]Y. Nishi and K. Iwata, Effects of Electron Beam Irradiation on Impact Value for Soda Glass. Materials Transactions. 2005, 46, 2241-2245

[5]H. Kobayashi, M. Salvia and Y. Nishi, Effects of Electron Beam Irradiation on Charpy Impact Value of GFRP. Journal of the Japan Institute of Metals. 2006, 70, 255-257

[6]K. Inoue, K. Iwata, T. Morishita, A. Tonegawa, M. Salvia and Y. Nishi, Enhancement of Charpy Impact Value by Electron Beam Irradiation of Carbon Fiber Reinforced Polymer. Journal of the Japan Institute of Metals. 2006, 70, 461-466

[7]K. Oguri, K. Fujita, M. Takahashi, Y. Omori, A. Tonegawa, N. Honda, M. Ochi, K. Takayama and Y. Nishi, Effects of Electron Beam Irradiation on Time to Clear Vision of Misted Dental Mirror Glass, Journal of the Materials Research. 1998, 13, 3368-3371

[8]K. Oguri, N. Iwataka, A. Tonegawa, Y. Hirose, K. Takayama and Y. Nishi, Misting-Free Diamond Surface Created by Sheet Electron Beam Irradiation. Journal of the Materials Research. 2001, 16, 553-557

[9]R. Sugiyama, K. Yamane, W. Nakao, K. Takahashi and K. Ando, Effect of difference in Crack-healing ability on fatigue behavior of Alumina/SiC composites. Journal of the Intelligent Materials System and Structure, 2008, 19, 411-415.

EFFECT OF Si_3N_4 ON THE INSTABILITY OF Li_2O-CONTAINING CELSIAN IN THE BAS/Si_3N_4 COMPOSITES

Kuo-Tong Lee
Department of Chemical Engr. and Department of Materials Engr., Mingchi University of Technology, Taishan, Taipei 24301, Taiwan, ROC

ABSTRACT

In-situ Si_3N_4 reinforced BAS composites are being explored for high-temperature structural applications. The main drawback to their use under thermal cycling conditions is the presence of the hexacelsian BAS phase, which will cause a volume change around 300°C. Effective additives, like Li_2O, can promote the polymorphic conversion of hexacelsian to celsian. XRD experiments show that in monolithic systems the Li_2O-containing celsian can exist at 1650°C; however, in BAS/Si_3N_4 composites it transforms into hexacelsian phase at 1600°C. The presence of Si_3N_4 could enhance the instability of the Li_2O-containing celsian phase above 1590°C, in which the hexacelsian phase becomes thermodynamically stable. In this study, the instability mechanism is indirectly verified as the doping effect of Si, which is coming from thermal decomposition of Si_3N_4 in the composites. It is well known that Si or SiO_2 is a network former, which could make the chain-like structure of celsian instable and enhance the formation of a layered hexacelsian structure.

INTRODUCTION

Barium aluminosilicate ($BaO \cdot Al_2O_3 \cdot 2SiO_2$, commonly referred to as BAS) is one of the most refractory glass-ceramics. For using as high-temperature structural materials, *in-situ* Si_3N_4 reinforced BAS composites have been explored having advantages of low cost, isotropic material properties and ease of manufacturing [1-5].

BAS exists primarily in three polymorphic forms: monoclinic celsian (referred to simply as celsian), hexagonal (known as hexacelsian) and orthorhombic (known as α-hexacelsian) [6]. Celsian is the stable phase from room temperature to 1590°C [7]. Above 1590°C, it undergoes a transformation to hexacelsian, which becomes stable up to 1760°C of the melting point [7]. However, hexacelsian always metastably exist at temperatures below 1590°C because the crystallization of hexacelsian is more ready than that of celsian and the reconstructive hexacelsian-to-celsian transformation is extremely sluggish [8]. The main drawback of using hexacelsian BAS for thermal cycling applications is that the reversible transformation of hexacelsian to orthorhombic polymorph around 300°C causes a ~3% volume change [3, 7]. On the other hand, celsian BAS does not undergo any phase transformation up to 1590°C and is a desirable polymorph.

The kinetics of the transformation of hexacelsian to celsian can be promoted by various methods, including: prolonged heat treatment at high temperatures but below 1590°C [9], adding seed crystals to the reaction system [10, 11], using mineralizers [12, 13] and forming a solid solution with $SrAl_2Si_2O_8$ [14]. The effectiveness of different mineralizers or seeds on the conversion using various precursors has been extensively investigated in the past [10-15]. The main objectives of this study are to examine the stability of the Li_2O-containing celsian above 1590°C in monolith BAS and to pursue an insight into the instability of the Li_2O-containing celsian in composite BAS/Si_3N_4 system by conducting some controlled experiments.

EXPERIMENTAL PROCEDURE

Hexacelsian BAS was synthesized according to the procedure reported earlier [16], prefiring $BaCO_3$ (Mallinckrodt AR having 99% purity) and SiO_2 (Nyacol 2034DI colloidal silica having 20 μm particle size) at 1200°C for 4 h to yield $BaSi_2O_5$ and then reacting $BaSi_2O_5$ with Al_2O_3 (Baikowski SM-8 having 99.99% purity and 0.15 μm particle size) at 1200°C for 4 h to yield hexacelsian BAS. The hexacelsian was then doped with 5 mol% Li$_2$O (Alfa Aesar's reagent grade) and heated at 1100°C for 4 h to synthesize celsian BAS. The celsian powder was used for several designed experiments to examine the effects on the instability of Li$_2$O-containing celsian in the composite BAS/Si$_3$N$_4$ system.

For composite syntheses, BaCO$_3$, SiO$_2$, Al$_2$O$_3$ and Si$_3$N$_4$ (Hermann Starch LC-12S, 94.5% α-Si$_3$N$_4$ and 5.5% β-Si$_3$N$_4$) were weighed in the ratio of 70 vol% Si$_3$N$_4$ and 30 vol% BAS, with or without the addition of mineralizers. Billets of the powder mixtures were formed by isostatic pressing in an elastomer mold at 138 MPa and sintered in a graphite resistance heated furnace. Sintering conditions were 1800°C for 3 h in a 0.1 MPa N$_2$ gas atmosphere with heating and cooling rate of 5°C/min.

Phases present in a sample after heat-treated were identified [17] by X-ray diffraction (XRD) employing Cu Kα radiation. The XRD patterns were recorded at room temperature using a step scan procedure (0.02° 2θ/step, count time 0.5 s). Microstructural analysis was performed using scanning electron microscopy (SEM).

RESULTS AND DISCUSSION

Synthesis of Li$_2$O-Containing Celsian BAS

X-ray diffraction patterns in Figure 1. demonstrate the effect of starting materials on the celsian formation with or without the additive Li$_2$O. Figure 1(a) shows that the conversion into celsian from pre-synthesized haxacelsian powder without Li$_2$O is completed when heated at 1530°C for 50 h. However, the same heat treatment but reaction originating from the raw materials, BaCO$_3$, Al$_2$O$_3$ and amorphous SiO$_2$ powders, results in a very different pattern of x-ray diffraction in which the primary phase is hexacelsian with insignificant amounts of BaAl$_2$O$_4$ and celsian phases, as shown in Figure 1(b). The slow kinetics of hexacelsian formation via the reaction between BaAl$_2$O$_4$ and SiO$_2$, reported by Planz and Muller-Hesse [18], and Lee and Aswath[16], could be responsible for the residual BaAl$_2$O$_4$ phase in the sample, even after heat treatment at 1530°C for 50 h. However, it is surprising that the formed hexacelsian only undergoes very little conversion into celsian at temperatures as high as 1530°C for 50 h, implying the presence of BaAl$_2$O$_4$ affects the hexacelsian-to-celsian transformation in BAS.

The other pair of x-ray diffraction patterns shown in Figure 1(c) and (d) also indicates that the use of a pre-synthesized hexacelsian as a starting material could enhance the kinetics of the polymorphic conversion. With the doping of 5 mol% Li$_2$O, the formation of celsian from hexacelsian powder was accomplished at the temperature as low as 900°C for 4 h (Figure 1(c)). The role of Li$^+$ on the kinetics of the phase conversion from hexacelsian to celsian BAS has been reported by many authors [15, 19-22]. However, when a mixture of BaCO$_3$, Al$_2$O$_3$, SiO$_2$ and Li$_2$O powders in a designed ratio was used as a starting material and heated at 1050°C for 4 h, the formation of celsian results with the presence of an appreciable amount of hexacelsian and some unknown phases, shown in Figure 1(d). The unknown peaks may correspond to some intermediate phase produced from Li$_2$O and the other constituents in the sample.

The Effectiveness of Mineralizers on the Celsian Formation in the 30% BAS-70% Si$_3$N$_4$ Composites

While in the last paragraph, the formations of celsian BAS are from solid-sate mixtures with or without additive Li$_2$O, in this paragraph the crystallizations of BAS are from liquid-state mixtures of Si$_3$N$_4$ and BAS containing different additives. Some works in BAS/Si$_3$N$_4$ composites [1-5] have shown that at 1800°C β-Si$_3$N$_4$ whiskers can be grown in-$situ$ from α-Si$_3$N$_4$ in the presence of liquid BAS, implying that the liquid phase of BAS has adequate wet ability and solubility with the Si$_3$N$_4$ starting particles. The liquid BAS phase is essential to ensure the progress of the transformation of α-Si$_3$N$_4$ to β-Si$_3$N$_4$. In previous studies, the methods to crystallize celsian BAS from an amorphous phase include: doping mineralizers like Li$_2$O and NaF, forming a solid solution with SrAl$_2$Si$_2$O$_8$, and adding a seed like ZrO_2. Therefore, four additives of Li_2O, NaF, SrF_2 and ZrO_2 are selected to study their effectiveness on the *in-situ* crystallization of celsian.

As shown in Fig. 2, XRD patterns of five batched samples all indicate that a complete transformation of α- to β-Si_3N_4 is achieved and that the melt of BAS re-crystallizes as hexacelsian phase after cooling below the melting point of 1760°C at a rate of 5°C/min. No celsian phase is detected in the 30% BAS-70% Si_3N_4 composites with or without the addition of mineralizers. The evaporation of Li^+ and Na^+ during liquid-phase sintering at 1800°C could be responsible for the lack of celsian formation. However, SrF_2 and ZrO_2 are expected to be still present in the sintered composites because ZrO_2 is a refractory phase having a melting point of 2700°C and a significant amount of SrF_2 left in the sintered sample, as shown in Fig. 2(d). Two possibilities could account for the inefficiency of mineralizers in promoting the celsian formation in the 30% BAS-70% Si_3N_4 composites: celsian might become instable at temperatures above 1590°C and then transform back to hexacelsian; Si_3N_4 or N_2 atmosphere might play a role on the transformation of celsian to hexacelsian. A serial of controlled experiments was designed and conducted as follows to figure them out.

The Meta-Stability of Li_2O-Containing Celsian at Temperatures above 1590°C

Lin and Foster [7] indicated that above 1590°C, celsian would undergo a transformation to hexacelsian and the formed hexacelsian could metastably exist at temperatures below 1590°C. X-ray diffraction patterns in Figure 3. show the pure celsian phase without containing Li_2O would convert to hexacelsian after heated at 1650°C. For 1 h at 1650°C, there is a lot of hexacelsian transformed from celsian (Figure 3(a)) and the conversion is almost completed for 3 h at 1650°C (Figure 3(b)).

Fig. 4(a) shows the pure celsian BAS was made from the mixture of hexacelsian and 5 mol% Li_2O heated at 1100°C for 4h. The Li_2O-containing celsian was then heated at 1650°C for 1 h, a temperature where celsian BAS becomes unstable [7]. As shown in Fig. 4(b), there is no reaction happened, implying that the additive Li_2O not only could promote the celsian formation from hexacelsian BAS below 1590°C, but could also stabilize the structure of celsian above 1590°C.

Comparison of Fig. 4(c) with 4(b) indicates that Si_3N_4 might play an important role in enhancing the instability of the Li_2O-containing celsian at temperatures above 1590°C. As shown in Fig. 4(c), when mixed with 70 vol% Si_3N_4 and heated at 1600°C for 1 h in a N_2 atmosphere, the Li_2O-containing celsian almost completely transform to hexacelsian. Apparently, the presence of Si_3N_4 makes the formation of Li_2O-containing celsian suppressed, but how does it work? It can be found that after heated at 1600°C for 4 h, the compact of the Li_2O-containing celsian and 70 vol% Si_3N_4 will lose 11.034% in weight. The mass loss is due to the vaporization from pyrolysis of Si_3N_4. Krishnarao et. al. [23] synthesized composite powders of $MoSi_2$ and SiC by reacting mixture powders of Mo-SiO_2-C and Mo-Si_3N_4-C, respectively, at 1300°C. They identified the formation mechanism of $MoSi_2$ as the SiO_2 vapor in the Mo-SiO_2-C system and Si vapor from thermal decomposition of Si_3N_4 in the Mo-Si_3N_4-C system reacting with Mo particle and forming bulk silicides [23]. Their experimental results indicate that using Si_3N_4 as a preform of reactant Si to produce $MoSi_2$ is similar to using SiO_2. In order to investigate the effect of Si vapor from Si_3N_4 on promoting the instability of the Li_2O-containing celsian at temperatures above 1590°C, the effect of SiO_2 is studied as follows.

Effect of SiO_2 on the Instability of Li_2O-Containing Celsian

When BAS is used as a matrix in Si_3N_4-bearing composites, it has been shown in Fig. 2, Fig. 4(c) and earlier studies [1, 3, 22] that the formation of celsian is suppressed and hexacelsian exists as a metastable phase below 1590°C. In this study, it is assumed that the dopant Si, produced from Si_3N_4, is responsible for the suppression effect of Si_3N_4 and the effectiveness of Si doped into BAS system is equivalent to that of SiO_2 [23]. Controlled experiments are conducted to study the amount effect of SiO_2 and the temperature effect of the SiO_2 doping, as shown in Fig. 5.

Li$_2$O-containg celsian powders doped with 5 mol% or 10 mol% of SiO$_2$ are respectively heated at 1600°C and 1650°C for 4 hr in atmospheres of N$_2$ gas or ambient stagnant air. The XRD experiments show the results in stagnant air (figures not shown) are same as those in N$_2$ atmosphere (Fig. 5), indicating that there is no effect of N$_2$ gas on the promoting instability of celsian. As shown in Fig. 5(a) and 5(b), when heated at 1600°C, Li$_2$O-containg celsian with the presence of SiO$_2$ of 5 mol% or 10 mol% still persistently exists. In Fig. 5(c) and 5(d), it is shown that the threshold temperature of the celsian-to-hexacelsian transformation is 1650°C and the extent of transformation is increasing with the doping amount of SiO$_2$. It is well known that Si or SiO$_2$ is a network former, which could make the crankshaft-like chain structure of the (Al,Si)O$_2$ tetrahedra in celsian instable and promote the layer structure of the (Al,Si)O$_2$ tetrahedra in hexacelsian formed. Comparison of 10 mol% of SiO$_2$ doped in Fig. 5(d) with 70 vol% of Si$_3$N$_4$ added in Fig. 4(c), insufficient doping amount for need could be explained the reason why the transformation is incomplete in Fig. 5(d).

In Figure 6., SEM micrograph shows the Li$_2$O-containing celsian mixed with 70 vol% Si$_3$N$_4$ has liquid phases presented after heated at temperatures above 1500°C for 4 h in a N$_2$ atmosphere. The containing of Li$_2$O in BAS could lower its melting temperature by about 260°C.

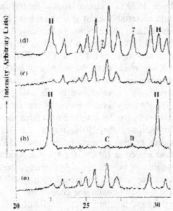

Figure 1. X-ray diffraction patterns show the effect of starting materials on the celsian formation with or without doping with Li$_2$O. (a) Complete celsian formed from hexacelsian powder heated at 1530°C for 50h; (b) Insignificant amount of celsian formed from a mixture of BaCO$_3$, Al$_2$O$_3$ and SiO$_2$ powders heated at 1530°C for 50h; (c) Complete celsian formed from hexacelsian powder with the presence of 5 mol% Li$_2$O heated at 900°C for 4h; (d) Incomplete celsian formed from a mixture of BaCO$_3$, Al$_2$O$_3$ and SiO$_2$ powders with 5 mol% Li$_2$O heated at 1050°C for 4h. (H) represents peaks from hexacelsian BAS, unmarked peaks and (C) celsian BAS. (B) BaAl$_2$O$_4$ and (?) unidentified phase.

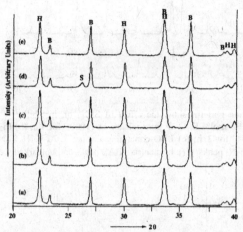

Figure 2. X-ray diffraction patterns of five 30vol% BAS-70 vol% Si$_3$N$_4$ composites heated at 1800°C for 3 h: (a) without additive, (b) with 5 mol% Li$_2$O, (c) with 10 mol% NaF, (d) with 50 mol% SrF$_2$ and (e) with 10 mol% ZrO$_2$. (B) represents β-Si$_3$N$_4$ peaks, (H) hexacelsian BAS peaks and (S) SrF$_2$ peaks.

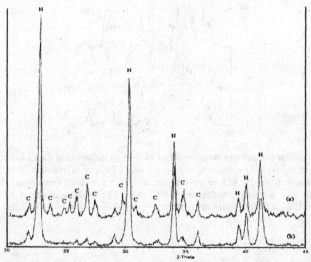

Figure 3. X-ray diffraction patterns show the instability of celsian without containing Li$_2$O when heated at 1650°C. Pure clesian heated (a) at 1650°C for 1 h and (b) at 1650°C for 3 h. (H): hexacelsian BAS peaks, unmarked peaks and (C): celsian BAS peaks.

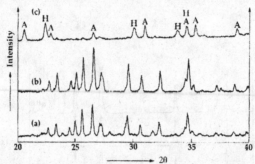

Figure 4. X-ray diffraction patterns show the effect of Si_3N_4 on instability of the Li_2O-containing celsian. (a) Hexacelsian + 5 mol% Li_2O → pure celsian, at 1100°C for 4 h; (b) Li_2O-containing celsian → no reaction, at 1650°C for 1 h; (c) Li_2O-containing celsian + 70 vol% Si_3N_4 → hexacelsian, at 1600°C for 1 h. (A): α-Si_3N_4 peaks, (H): hexacelsian BAS peaks and unmarked: celsian BAS peaks.

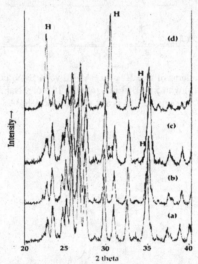

Figure 5. X-ray diffraction patterns show the effect of SiO_2 on instability of the Li_2O-containing celsian heated for 4 h in N_2. (a) Li_2O-containing celsian + 5 mol% SiO_2 → no reaction, at 1600°C; (b) Li_2O-containing celsian + 10 mol% SiO_2 → no reaction, at 1600°C; (c) Li_2O-containing celsian + 5 mol% SiO_2 → celsian + hexacelsian (a little), at 1650°C; (b) Li_2O-containing celsian + 10 mol% SiO_2 → celsian + hexacelsian (a major phase), at 1650°C.

Figure 6. SEM micrograph of the Li_2O-containing celsian mixed with 70 vol% Si_3N_4 and heated at (a) 1400°C for 4 h, (b)1500°C for 4h and (c)1600°C for 4h in a N_2 atmosphere.

SUMMARY
1. When *in-situ* 70 vol% Si_3N_4 reinforced BAS composites are synthesized, the formation of celsian is suppressed, even respectively with the effective mineralizers of Li_2O, NaF, SrF_2 and ZrO_2, and hexacelsian exists as a metastable phase below 1590°C.
2. The XRD experiments show that in monolithic system the Li_2O-containing celsian can persistent exist at 1650°C; however, in BAS/Si_3N_4 composite it transforms into hexacelsian phase at 1600°C. It is found that the presence of Si_3N_4 could enhance the instability of the Li_2O-containing celsian above 1590°C, in which the hexacelsian is a stable phase in thermodynamics.
3. Li_2O-containing celsian powders added with 5 or 10 mol% of SiO_2 and then heated at temperatures above 1600°C would become instable and transform into hexacelsian phase.
4. The instability mechanism of the Li_2O-containing celsian in BAS/Si_3N_4 composite possibly is the doping effect of Si, which is coming from thermal decomposition of Si_3N_4 in the composites.
5. The celsian BAS containing Li_2O would resist the transformation to hexacelsian at temperatures above 1590°C and would lower the melt point at about 1500°C, much lower than 1760°C of the celsian BAS without containing Li_2O.

REFERENCES
[1]K. K Richardson, D.W. Freitag and D.L. Hunn, Barium Alminosilicate Reinforced In-Stu with Silicon Nitride, *J. Am. Ceram. Soc.*, **78**[10], 2622 (1995).
[2]A. Bandyopadhyay, S.W. Quander, P.B. Aswath, D.W. Freitag, K.K Richardson and D.L. Hunn, Kinetics of In-situ α to β-Si_3N_4 Transformation in a Barium Aluminosilicate Matrix, *Scripta Metallurgica et Materialia*, **32**[9], 1417 (1995).
[3]A. Bandyopadhyay, P.B. Aswath, W.D. Porter and O.B. Cavin, The Low Temperature Hexagonal to Orthorhombic Transformation in Si_3N_4 Reinforced BAS Matrix Composites, *J. Mater. Res.*, **10**[5], 1256 (1995).
[4]S. W. Quander, A. Bandyopadhyay and P.B. Aswath, Synthesis and Properties of in situ Si_3N_4-Reinforced BaO-Al_2O_3-2SiO_2 Ceramic Matrix Composites, *J. Mater. Sci.*, **32**, 2021-29 (1997).
[5]F. Yu, C.R. Ortiz-Longo, K.W. White, *J. Mater. Sci.*, **34**, 2821 (1999).
[6]N. P. Bansal and M.J. Hyatt, Crystallization Kinetics of BaO-Al_2O_3-2SiO_2 Glasses, *J. Mater. Res.*, **4**[5], 1257-65 (1989).

[7]H. C. Lin and W.R. Foster, Studies in the System BaO-Al$_2$O$_3$-2SiO$_2$ I. The Polymorphism of Celsian, *Am. Mineral.*, **53**, 134-44 (1968).

[8]B. Yoshiki and K. Matsumoto, High-Temperature Modification of Barium Feldspar, *J. Am. Ceram. Soc.*, **34**[9] 283-6 (1951).

[9]K-T.Lee and P.B. Aswath, Enhanced Production of Celsian Barium Aluminosilicates by a Three-Step Firing Technique, *Material Chemistry and Physics*, **71**, 47 (2001).

[10]N. P. Bansal, Comment on Kinetics study on the Hexacelsian-Celsian Phase Transformation, *Material Science and Engineering A*, **342**, 23 (2003).

[11]C. Liu, S. Komarneni and R. Roy, Crystallization and Seeding Effect in BaAl$_2$Si$_2$O$_8$ Gels, Ceramic Matrix Composites, *J. Mater. Sci.*, **32**, 2021-29 (1997).

[12]K-T. Lee and P.B. Aswath, Kinetics of the hexacelsian to celsian transformation in barium aluminosilicates doped with CaO, *International Journal of Inorganic Materials*, **3**, 687 (2001).

[13]K-T. Lee and P.B. Aswath, Role of Mineralizers on the Hexacelsian to Celsian transformation in the Barium Aluminosilicate (BAS) System, *Material Science and Engineering A*, **352**, 1-7 (2003).

[14]N. P. Bansal and C.H. Drummond III, Kinetics of Hexacelsian-to-Celsian Phase Transformation in SrAl$_2$Si$_2$O$_8$, *J. Am. Ceram. Soc.*, **76**[5] 1321-24 (1993).

[15]M. Chen, W.E. Lee and P.F. James, Synthesis of Monoclinic Celsian Glass-Ceramic from Alkoxides, *J. Non-Cryst. Solids*, **147**, 532 (1992).

[16]K-T. Lee and P.B.Aswath, Synthesis of Hexacelsian Barium Aluminosilicate by Solid-State Process, *J. Am. Ceram. Soc.*, **83**[12], 352 (2000).

[17]JCPDS card: #9-250 for α-Si$_3$N$_4$, #9-259 for β-Si$_3$N$_4$, #12-726 for hexacelsian BAS, #18-153 for celsian BAS

[18]J. E. Planz and H. Muller-Hesse, Investigations on Solid-State Reactions in the System BaO-Al$_2$O$_3$-SiO$_2$. Part I: Solid-State Reactions in the Sub-System BaO-Al$_2$O$_3$, BaO-SiO$_2$ and Al$_2$O$_3$-SiO$_2$, *Ber. Dtsch. Keram. Ges.*, **38**, 440-450 (1961).

[19]J.C. Debsikdar, Sol-Gel Route to Celsian Ceeramic, *Ceram. Eng. Sci. Proc.*, **14**[1-2], 405 (1993).

[20]M.C. Guillem Villar, C. Guillem Monzonis and J.A. Navarro, Reactions between Kaolin and Barium Carbonate: Inflence of Mineralizers: 1. Qualitative Study, *Trans. J. Br. Ceram. Soc.*, **82**[2], 69-72 (1983).

[21]M.C. Guillem Villar, C. Guillem Monzonis and P.E. Lopez, Reactions between Kaolin and Barium Carbonate: Inflence of Mineralizers: 2. Qualitative Study, *Trans. J. Br. Ceram. Soc.*, **82**[6], 197-200 (1983).

[22]A. Bandyopadhyay and P.B. Aswath, A Phase Transformation Study in BAS-Si$_3$N$_4$ System, *J. Mater. Res.*, **10**[12], 3143 (1995).

[23]R.V. Krishnarao. V.V Ramarao and Y.R. Mahajan, J. Mater. Res., **12**[12], 3322 (1997).

Fatigue, Wear, and Creep

ROLLING CONTACT FATIGUE PROPERTIES AND FRACTURE RESISTANCE FOR SILICON NITRIDE CERAMICS WITH VARIOUS MICROSTRUCTURES

Hiroyuki Miyazaki, Wataru Kanematsu, Hideki Hyuga, Yu-ichi Yoshizawa, Kiyoshi Hirao and Tatsuki Ohji
National Institute of Advanced Industrial Science and Technology (AIST)
Anagahora 2266-98, Shimo-shidami, Moriyama-ku,
Nagoya 463-8560, Japan

ABSTRACT

In order to assess the usefulness of the indentation fracture (IF) method as a quick screening tool for bearing grade silicon nitrides (Si_3N_4), rolling-contact fatigue (RCF) performance was compared with both hardness and fracture resistance obtained by the IF technique using five Si_3N_4 ceramics with different microstructures. The RCF performance was evaluated by a ball-on-flat method with the load being increased in a stepwise manner. Attempts to correlate the fatigue properties directly with the hardness and fracture resistance were unsuccessful. A good correlation was obtained between the RCF performance and both mechanical properties when a lateral-crack chipping model was applied as the material removal process which limited the RCF life of the samples. However, the correlation was lost when the fracture toughness obtained by the single-edge precracked beam (SEPB) method was employed, indicating that the conventional long-crack toughness is inappropriate for analyzing the RCF performance of Si_3N_4 exhibiting a rising R-curve behavior. It was found that the IF method was suitable to rank the RCF performance by using the lateral-crack chipping model.

INTRODUCTION

Silicon nitride (Si_3N_4) ceramics have been successfully applied as ball bearings, and are now manufactured and used widely since they possess superior wear resistance and advantages such as light weight, high strength and toughness, good corrosion resistance and a low thermal expansion coefficient.[1] For such applications, evaluation of rolling-contact fatigue (RCF) performance is necessary. However, the RCF tests are not suitable means to discriminate among many bearing-quality Si_3N_4 which have been supplied in the market since they usually need long periods of time, ~ 20 days, depending on the loading conditions, etc.[2] Thus, a simple properties-based screening method is needed.

The fatigue properties are likely to correlate with the tolerance for localized damage. If the relation between the RCF performance and mechanical properties such as hardness and/or indentation fracture resistance can be revealed, a rough estimation of the former from the latter would become possible, which would be a convenient method for quick screening purposes since the indentation fracture (IF) technique is a quick and simple method.[3] However, there have been a only few studies on the relation between the RCF performance and both hardness and toughness. Burrier examined the RCF performance of 11 candidate monolithic silicon nitrides and ranked them with respect to their measured RCF lives.[2] By comparing the ranking with such properties as elastic modulus, hardness and toughness, he found a tendency that harder material exhibited better performance with a few exceptions, while the direct correlation between the RCF performance and fracture toughness was poor. By contrast, Xu et al. have shown that a function of the hardness and the indentation fracture resistance, which was derived from a lateral-crack chipping model,[4] was essential in analyzing abrasive machining behavior of silicon nitride since it corresponds to the microfracture processes.[5] Thus, it is reasonable to infer that the function of the hardness and indentation fracture resistance may be a useful indicator for the RCF performance of Si_3N_4 as well.

In this study, silicon nitrides with various microstructures from fine and uniform to coarse have been prepared and their RCF performances were measured using a ball-on-flat method in lubricant oil. The obtained results were compared directly with the hardness and fracture resistance determined by the indentation fracture (IF) method, as well as fracture toughness from the SEPB method. The ability of the function of both hardness and fracture toughness to rank the RCF performance was discussed in conjunction with their R-curve behavior, etc.

EXPERIMENTAL PROCEDURE
Materials

Five kinds of Si_3N_4 ceramics (listed in Table 1) with different microstructures were used in this study. One was prepared by the authors using small amounts of sintering additives (material G)[6] and four materials were obtained from commercial sources. Materials A and D1 were Si_3N_4 for ball bearings and materials C2 and F were Si_3N_4 for other aplications. Materials A and D1 were hot-isostatically pressed and material G was hot-pressed. Materials C2 and F were pressure-less sintered. The machined samples were polished and plasma etched in CF_4 gas before microstructural observation by scanning electron microscopy (SEM). Figure 1 shows microstructures of the F and G samples. It can be found that the F sample contained some coarse grains, whereas the grains of G sample were much smaller than that of F sample and the microstructure was more uniform. Samples A and D1 showed similar fine and uniform microstructures to that of the G sample, which was confirmed quantitatively by the average grain diameters listed in Table I. The microstructure of C2 sample resembled to that of the F sample.

Table 1   Sintering Methods and Grain Diameters of Si_3N_4 samples

Sample	A	D1	C2	F	G
Sintering Method	HIP	HIP	PLS	PLS	HP
Average Grain Diameter (μm)	0.33	0.27	0.77	0.59	0.35

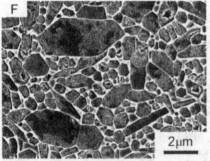
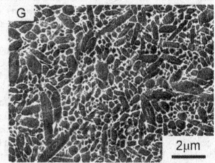

Figure 1.   SEM micrographs of the samples F and G.

Characterization of Toughness and Hardness

Vickers indentations were made on the polished surface with loads of 49, 98, 196, 294 and 490 N to vary the crack size over a broad range. The lengths of the impression diagonals, $2a$, and sizes of surface cracks, $2c$, were measured with a traveling microscope immediately after the unloading. The fracture resistance, K_R, was determined from the Miyoshi's equation as follows:[7,8]

$$K_R = 0.018(E/H_V)^{1/2} P c^{-3/2}$$ (1)

where E is Young's modulus and H_V is the Vickers hardness. P is the indentation load and c is the half-length of as-indented surface crack length.

For fracture toughness, K_{Ic}, measurement, rectangular specimens (4 mm in width x 3mm in breadth x ~40 mm in length) were machined from each sintered sample. The SEPB test was performed according to Japanese industrial standard (JIS) R 1607 with a pop-in crack depth of about 2 mm.[8]

Fatigue Test

The rolling-contact fatigue (RCF) tests were conducted by the ball-on-flat method. The schematic test configuration is shown in figure 2. Three Si_3N_4 balls which were made of material A were set on the Si_3N_4 sample plate. The balls were rotated by the main spindle at the speed of 1800 rpm in the lubricant oil, of which viscosity is ISO VG68, with the load being applied from bottom to up. Failure (surface peeling or spall) of the specimens was detected by an accelerometer mounted on the loading arm. Three to five specimens were tested for each sample. In these tests, the load was increased in a stepwise manner, 1, 2.5, 4, 5.5 kN to shorten the testing time. The number of contact for each step was 10^6.

In order to compare the results from the stepwise method with those from the conventional constant load method, the mean effective load, P_m, was calculated as follows,

$$P_m = \sqrt[3]{\frac{\sum_i P_i^3 n_i t_i}{\sum_i n_i t_i}}$$ (2)

where P_i is the load in N. n_i is the rotation speed in rpm and t_i is the testing time in minutes. Subscript i indicates the ith step, It is likely that the tolerance to rolling-contact fatigue is expressed well by the P_m times the number of contacts, N, since the degree of damage accumulated during the test inevitably depends not only on the load but also on the number of contacts, N. Thus, the value of NP_m was used to evaluate the RCF performance in this study.

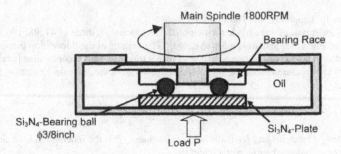

Figure 2. Schematic illustration of the ball-on-flat test method for the evaluation of the rolling contact fatigue performance.

RESULTS

Table II summarizes mechanical property data measured by the present authors for the five Si_3N_4 ceramics. The Young's modulus, E and hardness, H_V of G sample were highest among those samples, which was attributable to the small volume fraction of intergranular phase that is softer than pure Si_3N_4. By contrast, the fracture toughness for the G sample was lowest, which can be also explained by the small volume fraction of sintering additives as follows. The small volume fraction of intergranular phase in the G sample is expected to result in a rigid bonding between the grains which suppress crack bridging, whereas the bonding between grains in other samples should be weaker than that for the G sample since a sufficient amount of intergranular phase is present to allow for crack bridging to occur more frequently, leading to the higher fracture toughness. The Young's modulus and hardness of ball bearing grade samples (A, D1) were higher than those for other applications (C2, F). C2 sample showed the lowest Young's modulus and hardness, which may be due to the presence of a small volume fraction of pores. The fracture toughness of F sample was the highest, which can be explained by its coarse and elongated microstructure.[9,10] Sample C2 had the second lowest K_{Ic} despite its coarse microstructure, which may be attributable to the amount and kinds of additives, as well as the presence of a small volume fraction of pores.

Table II Mechanical properties of Si_3N_4 samples

Sample	A	D1	C2	F	G
Young's modulus (GPa)	305	309	289	295	315
Vicker's Hardness (GPa)*	15.1±0.2	15.7±0.2	12.4±0.2	13.8±0.2	16.6±0.2
Fracture toughness $(MPa \cdot m^{1/2})$	5.6±0.1	5.9±0.2	5.2±0.2	6.4±0.2	4.5±0.1

*Hardness was attained from the indentation with the load of 98 N.
Uncertainties are one standard deviation.

The ratio of as- indented crack length to the characteristic dimensions of the "plastic" impression, c/a, was larger than 2 in the range of the indentation load investigated for all samples, confirming that the cracks in this study were median-radial cracks.[11,12] Thus, equation (1) can be used for assessing the fracture resistance. The fracture resistances determined from the as-indented crack lengths are shown in figure 3 as a function of the indentation load. The fracture resistances of the C2 and G samples were almost the same and showed little dependence on the indentation load. In the case of the A and F samples, the increase in K_R with the load was more evident than that for C2 and G samples, indicating the presence of rising R-curve behavior for these materials. The D1 samples showed the highest fracture resistance among those samples, although its fracture toughness was the second highest as shown in table II. As compared with the fracture toughness measured by the SEPB method, the discrepancy between K_{Ic} and K_R was largest for sample F, which is due to the significant rising R-curve behavior caused by its coarse and elongated microstructure.[9,10]

Figure 4 shows the results of rolling-contact fatigue tests. The fatigue property was evaluated by the parameter NP_m which was determined from the mean effective load, P_m in kN times the number of contacts, N in millions. It is obvious that material A and D1 (bearing grade) showed superior NP_m values than those of other samples. Material C2 showed the smallest value, which was easily expected from both lowest Young's modulus and hardness, H_V values.

The relations between the fatigue performance and both H_V and K_{Ic} are presented in figures 5 and 6, respectively. In figure 5, there seems to be an increasing tendency of NP_m with an increase in hardness, but the correlation was not so good as displayed by the correlation coefficient, R^2 in the figure. The correlation was even worse when the NP_m was plotted against the fracture toughness obtained by the SEPB method (figure 6). No correlations were also found when the fracture resistances at various indentation loads were used as X axis in figure 6. It is apparent from those results that finding simple liner relations between the fatigue property and both K_R and H_V was difficult, which was consistent with the report by Burrier.[2]

DISCUSSION

The mechanism of material removal in abrasive machining of brittle ceramics has been well documented by Evans and Marshall who developed the lateral-crack chipping model.[4] They demonstrated that material removal rate, ΔV, caused by a passage of each grinding particle is related to the peak normal penetration forces, Fn, Vicker's hardness and fracture resistance as follows,[4]

$$\Delta V \propto Fn^{9/8}/(K_{Ic}^{1/2}H_V^{5/8}) \tag{3}$$

In this study, there were no grinding particles which were used in their model. However, it is reasonable to expect that a large piece of Si_3N_4 debris may be produced from the wear of the Si_3N_4 ball or substrate, which may act as a sharp indenter. If this is the case, the model is likely to be applicable to this study. Although it is observed that Hertzian cone-crack initiation and propagation resulted in large volume material removal under cyclic contact loading,[13] the quantitative relationship between the RCF performance and mechanical properties has not been established. For convenience' sake, Evans and Marshall's model was employed to correlate RCF properties with the hardness and fracture resistance. It should be noted that a speculative assumption was made that the probability of the generation of such a large piece of Si_3N_4 debris was equal for all the samples in applying the model to our case. The peak normal force, Fn, can

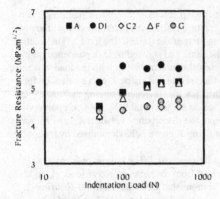

Figure 3. Fracture resistance determined by the IF method at various indentation loads. The size of almost all the error bars (± 1 standard deviation) was similar to the size of the symbols.

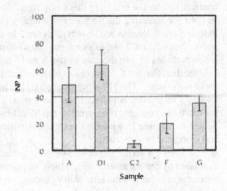

Figure 4. Results of the rolling contact fatigue test expressed by the parameter NP_m which was calculated from the mean effective load in kN times the number of contacts in millions.

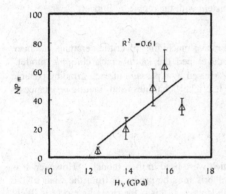

Figure 5. NP_m versus Vicker's hardness obtained at a load of 98 N. Error bars are ± 1 standard deviation.

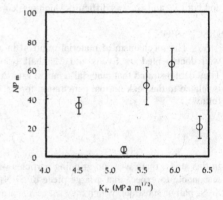

Figure 6. NP_m versus fracture toughness determined by the SEPB method. Error bars are ± 1 standard deviation.

be replaced by the mean effective load, P_m. Providing that the failure of the bearing system was detected when a certain amount of volume was removed, ΔV times the loading time or the number of contacts, N, should be constant regardless of the material used. Then, the next equation is obtained.

$$N^{8/9}P_m \propto K_{1c}{}^{4/9}H_V{}^{5/9} \qquad (4)$$

In this study, $N^{8/9}$ was approximated as N. In figure 7, NP_m is plotted against $K_{1c}{}^{4/9}H_V{}^{5/9}$. The data point of F sample clearly deviated from the best-fit line in figure 7, which caused the low correlation coefficient, R^2. By contrast, a good correlation was obtained when the fracture resistance, K_R, obtained by the IF method at the load of 98 N was applied instead of K_{1c} (figure 8). When figure 7 is compared with figure 8, the degree of shift in X axis for F sample was much larger than those of other samples. The large shift in X axis for F sample was originated from the discrepancy between the values of K_{1c} and K_R, which can be explained by the steep rising R-curve behavior for this sample. It is clear that the K_{1c} obtained at long crack length is inappropriate to analyze the fatigue performance with this model. By contrast, K_R obtained by the IF method at low load was preferable to correlate the fatigue properties with H_V and K_R. The crack depth of indentation at 98 N ranged between 130 ~ 150 μm, which was significantly smaller than that of SEPB, ~2 mm. The depth of spalls observed in the fracture surfaces of the damaged samples after the test was in the order of ~ 100 μm. It can be inferred that the size of cracks which might cause the spalls in damaged samples must be much closer to the depth of crack for the IF test at 98 N than that of SEPB. Therefore, the superiority of K_R from IF to K_{1c} from SEPB is reasonable in the analysis of the localized damage.

It seems that the rolling-contact fatigue life in this study was governed by the lateral fracture mechanism. However, the direct evidence of lateral cracks, which were induced by an indenter, could not be observed around the damaged portion of the samples. Some of the fracture origins of the damaged samples were peelings which had depths on the order of several μm. In this case, the lateral-crack chipping model was difficult to apply since the peeling was apparently caused by the grain dropping. Furthermore, the assumption of the equal probability of generation of a large piece of debris which acts as a sharp indenter was doubtful. This may be the reason why the best-fit line in figure 8 did not intercept the origin. The exponents for hardness and fracture resistance for the microfracture process may be different from those for the lateral crack chipping model in equation 3. Therefore, a new model for material removal in rolling contact fatigue life test is needed.

CONCLUSION

The present study investigated the effect of both the fracture resistance and hardness on the rolling contact fatigue (RCF) property. The RCF properties did not show the direct relation to the hardness as well as the fracture resistance. An indentation fracture model for material removal process in the rolling contact fatigue test was used to correlate the fatigue performance with fracture resistance and hardness of the materials. An agreement was obtained between the experimental results and the indentation model only when the fracture resistance associated with short cracks was used. This suggested that the conventional long-crack toughness was inappropriate for analyzing the rolling contact fatigue of Si_3N_4 exhibiting a rising R-curve behavior. From these results, it was concluded that the IF method was beneficial for ranking the RCF performance of silicon nitrides.

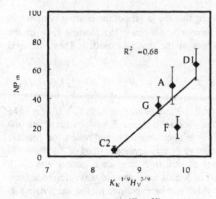 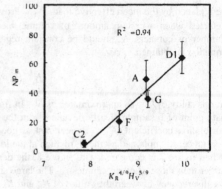

Figure 7. NP_m versus $K_{Ic}^{4/9}H_V^{5/9}$. K_{Ic} was obtained by the SEPB method. H_V was obtained at the load of 98 N. Error bars are± 1 standard deviation. The letter by every data point represents the sample name.

Figure 8. NP_m versus $K_R^{4/9}H_V^{5/9}$. K_R was obtained by the IF method at the load of 98 N. H_V was obtained at the load of 98 N. Error bars are± 1 standard deviation. The letter by every data point represents the sample name

ACKNOWLEDGMENT
This work has been supported by METI, Japan. as part of the international standardization project of test methods for rolling contact fatigue and fracture resistance of ceramics for ball bearings.

REFERENCES
[1] K. Komeya, Materials development and wear applications of Si_3N_4 ceramics, *Ceram. Trans.,* **133**. 3-16 (2002).
[2] H. I. Burrier. Optimizing the Structure and Properties of Silicon Nitride for Rolling Contact Bearing Performance, Tribol. Trans. **39** (2), 276-85 (1996).
[3] B. R. Lawn, A. G. Evans and D. B. Marshall, Elastic/Plastic Indentation Damage in Ceramics: The Median/Radial Crack system, *J. Am. Ceram. Soc.*, **63**, 574-81 (1980).
[4] A.G. Evans and D. B. Marshall: "Wear Mechanisms in Ceramics"; pp. 439-52 in *Fundamentals of Friction and Wear of Materials.* edited by D. A. Rigney. American Society for Metals, Metals Park, OH, (1981).
[5] H. H. K. Xu, S. Jahanmir, L. K. Ives, L. S. Job and K. T. Ritchie, Short-Crack Toughness and Abrasive Machining of Silicon Nitride, *J. Am. Ceram. Soc.*, **79**, 3055-64 (1996).
[6] H. Miyazaki. H. Hyuga, K. Hirao and T. Ohji, Comparison of Fracture Resistance Measured by IF Method and Fracture Toughness Determined by SEPB Technique Using Silicon Nitrides with Different Microstructures, *J. Eur. Soc. Ceram.*, **27**, 2347-54 (2007).
[7] T. Miyoshi, N. Sagawa and T. Sasa, Study of Evaluation for Fracture Toughness of Structural Ceramics, *J. Jpn. Soc. Mech. Eng.*, **A, 51**, 2489-97 (1985).

[8]JIS R 1607, Testing Methods for Fracture Toughness of Fine Ceramics," Japanese Industrial Standard, (1995).
[9]P. F. Becher, Microstructural Design of Toughened Ceramics, *J. Am. Ceram. Soc.*, **74**, 255-69 (1991).
[10]R. W. Steinbrech, Toughening Mechanisms for Ceramic Materials, *J. Eur. Ceram. Soc.*, **10**, 131-42 (1992).
[11]T. Lube, Indentation Crack Profiles in Silicon Nitride, *J. Eur. Ceram. Soc.*, **21**, 211-18 (2001).
[12]D. B. Marshall, Controlled Flaws in Ceramics: A Comparison of Knoop and Vickers Indentation, *J. Am. Ceram. Soc.*, **66**, 127-31 (1983).
[13]Z. Chen, J. C. Cuneo, J. J. Mecholsky and S. Hu, Damage Processes in Si_3N_4 Bearing Material Under Contact Loading, *Wear*, **198**, 197-207 (1996).

FRETTING FATIGUE OF ENGINEERING CERAMICS

Thomas Schalk, Karl-Heinz Lang, Detlef Löhe
University of Karlsruhe (TH), Institute of Materials Science and Engineering I
Kaiserstr. 12, D-76131 Karlsruhe, Germany

Keywords: Fretting fatigue, alumina, silicon nitride, cyclic fatigue, four-point-bending

ABSTRACT

Fretting fatigue appears in the contact zone between two components which are exposed to cyclic loading and where relative movement occurs even if the motion happens with very small stroke amplitudes in the range of few micrometers. In machined components of equipment, forces are often transmitted by form-fitting connections. During loading at least microsliding in the interfaces of such connections appears. In the contact zone a combination of normal- and shear stresses arise from the superposition of stresses resulting from Hertzian stresses and shearing due to the relative motion of the surfaces. In many cases this complex multi-axial loading environment determines the lifetime of the components.

Fretting fatigue of engineering ceramics has yet to be investigated sufficiently. Therefore, a new test rig to investigate fretting fatigue using four point bending specimens made of engineering ceramics was designed and built within the Collaborative Research Center 483: "High performance sliding and friction systems based on advanced ceramics" funded by the Deutsche Forschungsgemeinschaft (DFG). This test rig generates fretting fatigue loading through the superposition of cyclic four point bending and local friction loading. The friction loading is achieved by the relative motion of a well-defined fretting pad which is pressed on the surface of the four point bending specimen and moved by an electrodynamic actuator. The fretting pad oscillates on the specimen surface under tensile stress.

The results of the experiments illustrate the influence of different parameters of the fretting fatigue loading on the lifetime behavior of alumina and silicon nitride and the surface damage introduced by fretting fatigue.

INTRODUCTION

Fretting fatigue occurs if mechanical base load is superimposed with tribological loading. In the contact zone sliding friction may appear because of many different reasons. Fretting fatigue arises according to the following criteria [1]:

⇒ contact between the fretting fatigue bodies
⇒ relative movement between base body and antibody
⇒ cyclic mechanical loading which is able to induce fatigue

Though, fretting fatigue can even occur if the contact bodies only show different elastic material characteristics. This circumstance is exemplified in [2] by Hills and Nowell. In the field of aerospace numerous studies deal with fretting fatigue [2] – [12]. Special attention is given to the joint between blade and disk of a turbine, e.g. [13]. In these studies there is a focus on the material Ti-6Al-4V. A better understanding of the damaging processes induced by fretting fatigue contributes to an improved reliability and because of this a reduction of costs through more precise life-time predictions can be reached.

STATE OF TECHNOLOGY

Fretting fatigue of engineering ceramics is a research area that has not been investigated sufficiently so far, whereas for metallic materials many research reports and a relatively good database exist (e.g. [2]). Silicon nitride (Si_3N_4) is the only engineering ceramic material that was investigated under fretting fatigue conditions. Most of the papers related to this material are publicized Y. Mutoh et al. and M. Okane et al. [14]-[15]. An elementary publication with regard to the different procedures for the investigation of fretting fatigue have been assembled by R. B. Waterhouse [16]. In addition there are a number of publications dealing with the phenomenon of fretting fatigue in common. But these ones have a strong focus on metallic material. Many different models and methods to investigate fretting fatigue are described, giving a good survey of the actual state of research and technology on this issue.

An example where fretting fatigue is responsible for the development of damage is rolling of wires and bands. Here an oscillating relative movement between the forming rollers and the rolled material occurs. This is described e.g. in [17]. Figure 1 illustrates the rolling procedure and conditions in the contact zone schematically. In the following a rotating point of the surface of one of the rollers is considered. In this point cyclic loading arises due to rotation and the plastic deformation of the wire. Hence, the loading in this point is a combination of Hertzian stresses superposed by a complex tribological shear loading.

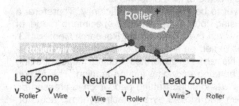

Lag Zone   Neutral Point   Lead Zone

$V_{Roller} > V_{Wire}$ $V_{Wire} = V_{Roller}$ $V_{Wire} > V_{Roller}$

Figure 1: Conditions in the contact zone during rolling of wires

In the contact area an advance zone and a lag zone are differentiated which develop in front of and after the neutral point respectively. By reason of the different thicknesses of the wire and the resulting different velocities between wire and rollers in the contact zone, friction forces and relative movements in driving direction and opposite to the driving direction exist.

Herefrom and because of the different elastic properties between the wire and the rollers and the shearing strain aroused by the extrusion process a fretting fatigue exposure of the rollers is the result. This exposure is able to initate viable fretting fatigue cracks which can lead to failure of the rollers.

BASIC CHARACTERIZATION

Many efforts to determine the mechanical characteristics of different ceramic materials have been made. Within the framework of the collaborative research centre 483 supported by the DFG ceramic material like alumina, silicon nitride, silicon carbide and sialon have been investigated all-embracing [19, 22, 23]. This means that there is a very good database on which the fretting fatigue experiments are built up. The results of the four-point-bending tests are the basis for the determination of the fretting fatigue influence on the life-time of cyclic stressed ceramic assemblies [18, 20]. In practice the four-point-bending test for determination of the mechanic strength characteristic of engineering ceramics has been established [19]. Here a specimen with the shape of a bar (3x4x45 mm^3) is loaded either dynamical or cyclical. Applying four-point-bending a constant bending moment arises in a relatively large area of the specimen. Isothermal material testing experiments have been carried out at room temperature and at high temperatures up to 900 °C. It is evident that cyclic loaded ceramics with glass phases on the grain boundaries are tending to show fatigue effects.

MATERIAL AND EXPERIMENTAL METHODS

Two different ceramic materials – alumina and silicon nitride - have been tested. The tested alumina material was produced by FRIATEC (Mannheim, Germany) with the quality F99,7. The material shows a low percentage of glass phase less than 0.3 mass percent. Fatigue tests have shown that this material exhibits a distinct cyclic fatigue effect in spite of the marginal content of glass phase.

The other tested material is silicon nitride type SL 200 from CERAMTEC (Plochingen, Germany). The material has a high resistance against thermal shock and a relatively high fracture toughness of about 4.9 MPa m$^{1/2}$. The material contains about 12 ma.-% glass phase between the grains. Examinations according to the fatigue behaviour have shown that cyclic fatigue effects appear at this material.

To realize the experiments existing four-point-bending test rigs were modified. The dimensions of the base load rig are appropriate to DIN 843-1 [19] as shown in Figure 2. So, the results of the fretting fatigue experiments can be compared with the results of the four-point-bending tests without any fretting loading from [18, 20, 22, 23].

To put fretting fatigue loading into practice a fretting pad is placed on the failure critical tension loaded surface of the bending specimen using a spring to reach a defined contact force. The oscillating movement of the fretting pad is realized by an electrodynamic actuator system that transfuses the friction forces via a system of levers. This actuator enables very small stroke amplitudes of the fretting pad less than 50 µm. The contact pressing force and the friction force are measured by load cells. The four-point-bending base load is generated by an electrodynamic testing machine with a loading appropriate to DIN-EN standard 843-1 [19].

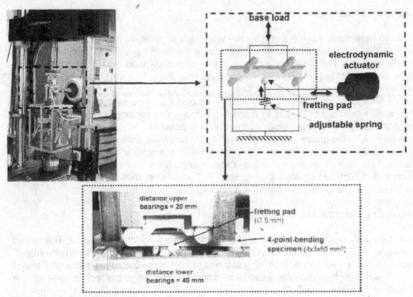

Figure 2: fretting fatigue test rig

RESULTS

First test series to check the residual strength of the specimens were carried out at room temperature in environmental atmosphere. The results introduced in [18, 20, 22, 23] serve as a benchmark for the fretting fatigue experiments with alumina and silicon nitride specimens. In these tests the specimens were clamped in the four-point bending rig without a cyclic four-point-bending base load in order to lock the specimens into position. Only the fretting pad oscillates on the surface of the specimens. This testing procedure was chosen to resolve the influence of the "fretting notch" aroused by the movement of the fretting pad and resulting from wear processes. Scanning electron microscope examinations at silicon nitride specimens have shown that damage develops on the fretted surface after a short period of time. Figure 3 shows the surface after 500 cycles of fretting. The arrows indicate the moving direction of the fretting pad.

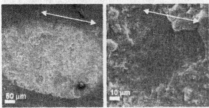

Figure 3: Surface after 500 cycles, p_{max}= 900 MPa; s_a= 100µm; f= 40Hz

The fretted surface appears rougher than the surrounding grinded surface of the specimen. First cracks are observed on the ground of the fretting notch. The cracks are growing with increasing duration of the experiment and small plates on the surface are quarried out. Figure 4 shows different crack patterns after 100,000 cycles of fretting. It can be seen clearly that there are areas where the cracks are orientated perpendicular to the moving direction of the fretting pad (Figure 4a). Other areas where plates are removed show the structure of the silicon nitride material. The major course of the crack at the fracture surface is intergranular (Figure 4b).

Figure 4: Surface after 100,000 cycles, p_{max}= 900 MPa; s_a= 100µm; f= 40Hz

With scanning electron microscope examinations alumina specimens have shown that disruption on the fretted surface appears after a short period of time, too. Figure 5 shows the surface after 1,000 cycles of fretting. The damage has another appearance than with silicon nitride specimens. Here severeal cracks appear on the surface which are orientated perpendicular to the moving direction of the fretting pad. The white arrows are showing the moving direction of the fretting pad.

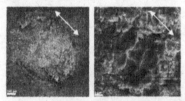

Figure 5: Surface after 1,000 cycles, p_{max}= 2911 MPa; s_a= 100µm; f= 40Hz

The cracks grow with increasing duration of the experiment and coalescence to a crack network. Figure 6 shows that after 5,000 cycles the fretting induced crack network can already be seen clearly. Further examinations will have to show in which depth the crack network has grown.

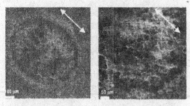

Figure 6: Surface after 5,000 cycles, p_{max}= 2911 MPa; s_a= 100µm; f= 40Hz

Subsequently the statically stressed specimens were claimed within a dynamic four-point-bending test with a load velocity of 1,000 MPa/s till fracture to determine the residual strength. Because of the local damage it is assumed that the residual material strength is reduced increasingly with an increasing number of fretting cycles.

The results of the residual strength measurements of silicon nitride in Figure 7 demonstrate that the static-fretting stressed specimens have a lower residual strength compared to the results from the inert strength tests. Figure 7 shows that the residual strength is independent of the number of fretting cycles. An analysis of the location from which rupture emanates has shown that some specimens were not fractured in the area of the fretting-notch but at macroscopical unaffected locations.

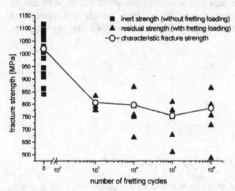

Figure 7: Residual strength of silicon nitride

In Figure 8 the results of the residual strength of tests with alumina specimens are illustrated. These experiments have shown that the static-fretting stressed specimens have – independent of the number of fretting cycles – virtually the same inert strength as non-fretted specimens.

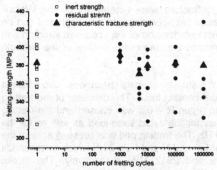

Figure 8: Residual strength of alumina

An analysis of the location from which rupture emanates has shown that some specimens were not fractured in the area of the fretting-notch but at macroscopical undamaged locations.

Then Wöhler diagrams at a stress ratio of R = 0.5 were created. Fretting fatigue test series were compared with a test series under four-point-bending without fretting conditions running at the same temperature and stress ratio conditions. To realize the fretting fatigue loading as fretting pad a ball was pressed against the surface of the specimens with a constant contact force.

In the test series carried out with silicon nitride the contact force was kept constant at F_N = 10N. The fretting pad was made of silicon nitride, too.The diameter of the ball varied between 5 and 15 mm, which corresponds to maximum H ertzian stresses in the contact area of p_{max} = 2,053 MPa and 987 MPa respectively. The testing frequency of the cyclic base load as well as of the oscillatory movement of the fretting pad was 40 Hz. The stroke amplitude of the fretting pad was chosen at s_a = 100 µm. This Wöhler diagram is shown in Figure 9.

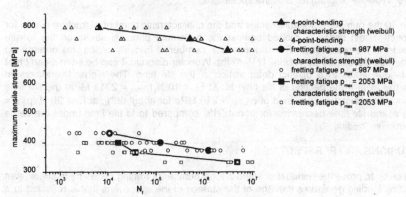

Figure 9: Woehler diagram for silicon nitride specimens

The plotted characteristic numbers of cycles to fracture were determined by the Weibull method [21]. Each point represents a test series of ten specimens. In the diagram it can be seen clearly that fretting fatigue leads to a distinct deterioration of the specimen strength and life time. The higher the maximum Hertzian stresses the clearer the decline of the strength and life time is.

Figure 10 shows the results of the fretting fatigue tests with alumina specimens. Two different test series were carried out under variation of the contact force. The diameter of the fretting pad was kept constant at 5 mm. As stress ratio again R = 0.5 was chosen and the fretting pad was made of alumina. The testing frequency for the cyclic base load as well as for the oscillatory movement of the fretting pad was 40 Hz. The fretting pad was pressed against the surface of the specimens with a contact force of F_N = 10 N and F_N = 20 N which corresponds to maximum Hertzian stresses of 2311 MPa and 2912 MPa respectively. The stroke amplitude of the fretting pad was chosen at s_a= 100 µm.

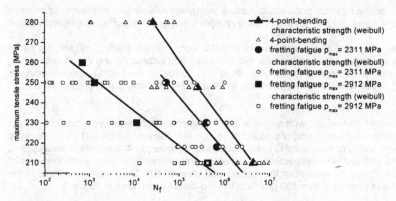

Figure 10: Woehler diagram for alumina specimens

In Figure 10 the number of fracture cycles and the characteristic number of fracture cycles for the respective test series is plotted against the maximum principal stress on the tensile loaded side of the specimens. The characteristic number of fracture cycles was determined by the method developed by Weibull [21]. In the Woehler diagram it can be seen clearly that fretting fatigue leads to a distinct deterioration of the life time. The higher the maximum stresses the clearer the decline of life time is. At F_N = 10 N (p_{max} = 2311 MPa) the life time decreases at a maximum base load of $\sigma_{R,max}$ = 210 MPa for about 80%, at F_N = 20 N (p_{max} = 2912 MPa) the life time decreases for about 91% compared to to life time under the same maximum base loading.

CONCLUSIONS AND PERSPECTIVE

Tests in order to prove the residual strength after a number of fretting cycles have shown that pure fretting loading generates damage of the surface of the specimens that is reflected in a lower residual strength at silicon nitride but not at alumina specimens. The superposition of cyclic four-point-bending and the oscillating movement of the fretting pad have shown that fretting fatigue is able to reduce fatigue strength and life time of ceramic specimens extensively. The experiments have shown that silicon nitride is more sensitive to fretting fatigue than alumina. There is a reduction in lifetime with alumina but a remarkable reduction of strength and lifetime with silicon nitride specimens under fretting conditions. Increasing maximum Hertzian stresses produce an increasing damage of the specimens resulting in lower fatigue strength and a shortened life time at both tested ceramics. Hence, lifetime reduction conditional on fretting fatigue has to be considered for dimensioning.

For further examinations the distribution of the cracks on the surface and in the depth of the specimens has to be analyzed. Experiments according to isothermal high-temperature exposure up to 900° C during fretting fatigue tests are in progress.

ACKNOWLEDGEMENTS

The authors thank the Deutsche Forschungsgemeinschaft (DFG) for financial support within the frame of the Collaborative Research Centre 483 „High performance sliding and friction systems based on advanced ceramics".

LITERATURE

[1] F.C. Neuner, Untersuchung von mikrostrukturellen Einflussgrößen auf das Reibermüdungsverhalten und auf die Schädigungsmechanismen von Eisenbasislegierungen, Dissertation Universität Erlangen 2006

[2] D.A. Hills, D. Nowell: Mechanics of Fretting Fatigue, Kluwer Academic Publishers, (1994).

[3] H. Helmi Attia, R.B. Waterhouse: Standardization of Fretting Fatigue Test Methods and Equipment, ASTM STP 1159, 1992

[4] T.N. Farris, H. Murthy, P.T. Rajeev: Fretting Fatigue of Ti6Al4V / Ti6Al4V and Ti6Al4V / In718 subjected to blade/disk contact loading, Journal of Fatigue, 2001, 9. 2153-2160

[5] S. Shirah, Y. Mutoh, K. Nagata: Fretting Fatigue Behaviour of Ti-6Al-4V Alloy and Structural Steel in Very High Cycle Regime, University of Agricultural Sciences, 2001, p. 295-302

[6] L.J. Fellows, D. Nowell, D.A. Hills: On the Initiation of Fretting Fatigue Cracks, Wear,205(1), 1997, p. 120-129

[7] Y. Mutoh, K. Tanaka: Fretting Fatigue in several steels and a cast iron, Wear,125(11), 1988, p. 175-191

[8] F.C. Neuner, R. Nuetzel, H.-W. Hoeppel: Fretting fatigue of carbon steels in the cycle fatigue regime, MP Materialprüfung 46 (7-8), 2004, p 379-383

[9] Alisha Hutson, Ted Nicholas, Reji John: Fretting fatigue crack analysis in Ti-6Al-4V, International Journal of Fatigue (27), 2005, p. 1582-1589

[10] H. Lee, S. Mall: Fretting behavior of shot peened Ti-6Al-4V under slip controlled mode, Wear (260), 2006, p.642-651

[11] Alisha Hutson, Ted Nicholas, Reji John: Fretting fatigue of Ti-6Al-4V under flat-onflat contact, International Journal of Fatigue (21), 1999, p.663-669

[12] David W. Hoeppner, V. Chandrasekaran, Charles B. Elliott: Fretting Fatigue : Current Technology and Practices, ASTM, 2000

[13] N.K. Arakere, G. Swanson, Fretting Stresses In Single Crystal Superalloy Turbine Blade Attachments, Journal of Tribology,Vol.123, p. 413-423, 04/2001

14   M. Okane, T. Satoh: Y. Mutoh, S. Suzuki: Fretting Fatigue behaviour of silicon nitride. Fretting Fatigue, Mechanical Publications, London , ESIS 18, 393-371, (1994).

15   Y. Mutoh: Mechanisms of fretting fatigue, Review in JSME International Journal 38(4), 405-415, (1981).

16   R.B. Waterhouse: Fretting Fatigue. Applied Science Publishers Ltd. London, (1981).

17   B. Avitzur: Friction during Metal Forming. In: P.J. Blau Ed.: Friction, Lubrication and Wear Technology, ASM Handbook Vol 18, ASM International (1992).

18   Th. Schalk, Th. Schwind, E. Kerscher, K.-H. Lang, D. Loehe: Tagungsband zum 3. Statuskolloquium des Sonderforschungsbereich 483, Thermische, mechanische und Reibermüdung ingenieurkeramischer Werkstoffe, 2007

19   DIN EN 843-1: Hochleistungskeramik - Mechanische Eigenschaften monolithischer Keramik bei Raumtemperatur - Teil 1: Bestimmung der Biegefestigkeit, Beuth-Verlag, Ausgabe: 2007-03

20   T. Lube, R. Danzer: "An ESIS Testing Program for a Silicon Nitride Reference Material", in: A. Neimitz, I.V. Rokach, D. Kocanda, K. Golos (Eds.), Fracture Beyond 2000 - Proc. of ECF 14, Vol. II, EMAS Publications, Sheffield, 2002, 401-408

21   W. Weibull: A Statistical Theory of the Strenght of Materials, Generalstabens Litografiska Anstalts Foerlag, Stockholm, 1939

22   T.Schalk, T.Schwind, K.-H.Lang, E.Kerscher, D.Löhe: Thermische , mechanische und Reibermüdung ingenieurkeramischer Werkstoffe, Sonderforschungsbereich 483, Tagungsband zum 3. Statuskolloquium am 18. Oktober 2007

23   Rachid Nejma: Verformungs- und Versagungsverhalten von Aluminiumoxidkeramik unter isothermer und thermisch-mechanischer Ermüdungsbeanspruchung, Dissertation Universitaet Karlsruhe (TH), 2007

INVESTIGATION INTO CYCLIC FREQUENCY EFFECTS ON FATIGUE BEHAVIOR OF AN OXIDE/OXIDE COMPOSITE

Shankar Mall
Department of Aeronautics and Astronautics
Air Force Institute of Technology
Wright-Patterson AFB, OH 45433-7765, USA

Joon-Mo Ahn
Technical Development Center
Agency for Defense Development
P.O. 35-5, Yuseong, Taejeon, Korea, 305-600

ABSTRACT

An oxide-oxide CMC with no interface, Nextel720™/alumina, was investigated under tension-tension fatigue condition at three frequencies (1, 100 and 900 Hz) to characterize the effects of cyclic frequency at room temperature. Cycles to failure at a stress level increased considerably with increase of the cyclic frequency from 100 to 900 Hz, but not much between 1 and 100 Hz. Local fiber/matrix interfacial bonding due to the frictional heating at the highest frequency of 900 Hz was observed during cycling. This may have caused an increase in the fatigue life/strength of the tested CMC system at the highest frequency of 900 Hz.

INTRODUCTION

Gas turbine engine community is exploring oxide/oxide and other class of ceramic matrix composites (CMCs) for their possible applications[1-4]. Since CMCs in these applications could be subjected to cyclic loading condition, their fatigue behavior need to be characterized. Fatigue behavior is influenced by the cyclic frequency besides many other factors/parameters. Fatigue behavior could become complex in CMCs relative to homogeneous materials due to the presence of fiber/matrix interface. Therefore the effects of high frequency on the fatigue behavior oxide/oxide CMCs is needed especially if these are employed in gas turbine engines. This study is focused in this direction and it involved the laboratory room temperature environment only.

MATERIAL

The oxide/oxide CMC studied was manufactured by COI Ceramics (San Diego, CA). The fibers were Nextel 720 fibers (3M) in an 8HSW, 0°/90° woven layers, with a density of ~2.78 g/cm³. Nextel720™ fiber is a meta-stable mullite having the chemical composition: Al_2O_3 of 85% (by weight) and SiO_2 of 15% (by weight). This mullite fabric preform was infiltrated with the matrix precursor of alumina by a sol-gel process. Nextel720™/alumina CMC was made by "vacuum bag" drying under low pressure and low temperature, and then followed by pressureless sintering technique. Fiber volume fraction was about 44%. Matrix porosity was about 24%, and such porosity level renders the matrix sufficiently weak and gives the composite excellent damage tolerance.

SPECIMEN

The CMC plate was cut into dog bone shaped specimens using a water jet. Test specimens for fatigue test had thickness, width and gage length of 2.8, 10.2, 30 mm, respectively, and total length of 150 mm for 1 Hz tests, and 2.8, 6.4, 20 mm for 100 and 900 Hz tests with total length of 64 mm. Fiberglass tabs were attached in the grip portion of each specimen to prevent the sliding of specimen.

TEST PROCEDURES

All tests were conducted at room temperature in air under ambient laboratory condition. Three cyclic frequencies were used; 1, 100 and 900 Hz. Fatigue tests were run at 1 Hz for maximum of 10^5 cycles, and at 100 and 900 Hz for maximum of 10^8 cycles, if specimen did not fail. These were the run-out limit of fatigue tests in this study. Fatigue tests at 1 Hz were conducted on a standard servo-hydraulic test machine. Fatigue tests at 100 and 900 Hz were conducted on a servo-hydraulic high frequency fatigue test machine (25 kN, 1000 Hz High-Cycle/Frequency Fatigue Test System, MTS Corp.). All fatigue tests were run under the load control mode with a stress ratio (minimum load/maximum load) of 0.05. Strains were not measured during fatigue test due to vibration and stability of extensometer at cyclic frequency of 100 Hz or higher. The temperature increase on the surface of specimens during the fatigue test was measured using an infrared (IR) camera. Fracture surfaces of failed specimens were gold coated and examined using a Scanning Electron Microscope (SEM).

RESULTS AND DISCUSSIONS

Fatigue

The S-N curves (i.e. the applied maximum stress versus cycles to failure relationships) of the tested Nextel720™/alumina CMC for three different frequencies of 1, 100 and 900 Hz are shown in Figure 1. The average ultimate tensile strength (UTS) is also shown in this figure. This clearly shows that cyclic frequency has effect on the fatigue response. The fatigue life at a stress level increased with increase in the cyclic frequency from 1 Hz to 900 Hz. For example, cycles to failure at 1, 100 and 900 Hz were about 600, 6,000 and greater than 1×10^8 at the applied maximum stress of 120 MPa, respectively. Further, it is to be noted that the slope of these S-N relationships is decreasing with increase in the cyclic frequency from 1 Hz to 100 Hz. In other words, the effect of frequency is diminishing at higher stress levels, and it is almost negligible at stress levels greater than ~130 MPa between 1 Hz and 100 Hz. On the other hand, the fatigue strength at 10^7 cycles at 100 and 900 Hz are about 105 and 125 MPa, respectively. Overall, results in Figure 1 indicate that cycles to failure increased with increasing cyclic frequency relatively much more from 100 to 900 Hz, but not much between 1 and 100 Hz.

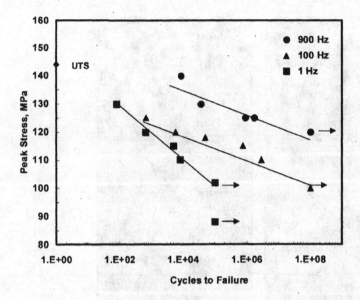

Figure 1. S-N relationships at three frequencies of 1, 100, and 900 Hz

There was increase in temperature on the specimens' surface during fatigue. Increase in surface temperature of specimens tested at 1 and 100 Hz was about 5°C during the fatigue test. However, surface temperature of specimens tested at 900 Hz increased with accumulating fatigue cycles, and this increase was more than 70°C before its failure. In general, temperature increase in CMCs during fatigue is attributed to the internal friction caused by the relative motion between constituents of the composite. Since the frictional heating in the composite having weak interface is generally dependent on the matrix crack density, and the number and length of debonded fiber/matrix interfaces, therefore, the temperature evolution suggest that the tested system had similar damage state and its progression during cycling at two lower frequencies of 1 and 100 Hz while it was different at the highest frequency of 900 Hz.

DAMAGE MECHANISMS
Microphotographs of fractured surfaces at the lowest and highest frequencies of this study, i.e. 1 and 900 Hz are compared in Figure 2. There was practically no difference in the damage mechanisms between 1 and 100 Hz, and hence 100 Hz case is not shown for the sake of brevity. There were differences on the fracture surface, which can be attributed to test frequency effects. Fracture surface in the matrix region of specimen tested at the frequency of 1 Hz showed only a few large matrix cracks and debris, while its counterpart at the frequency of 900 Hz showed relatively more amount of smaller matrix cracks and debris, Figures 2(a) and (b). On the other

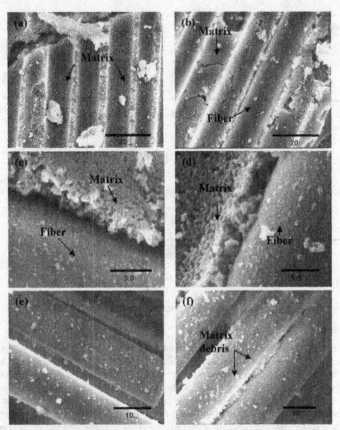

Figure 2. Microphotographs of Nextel^{TM}720/alumina CMCs at two frequencies; (a) matrix region at 1 Hz, (b) matrix region at 900 Hz, (c) interface between fiber and matrix at 1 Hz, and (d) interface between fiber and matrix at 900 Hz, (e) surface of fibers at 1 Hz and (e) surface of fibers at 900 Hz

hand, Figures 2(c) and (d) show the close up of fiber/matrix interfacial region from specimens tested at 1 and 900 Hz, respectively. This region in a specimen tested at low frequency of 1 Hz was relatively smoother that easily separated during fatigue test, and it is a typical feature of weak fiber/matrix interface. On the other hand, its counterpart at high frequency of 900 Hz showed quite rough region showing locally bonded fiber/matrix interface. This suggests that there was relatively more bonding strength at higher frequency of 900 Hz than at lower frequencies tests of 1 and 100 Hz. This bonding between matrix and fiber must have developed during cycling at the higher frequency of 900 Hz. This is also evident on the fibers which were separated from the matrix as shown in Figures 2(e) and (f). Fibers from specimen tested at the lower frequencies of 1 and 100 Hz had relatively smooth surface and they were separated individually in clean manner, while their counterparts at 900 Hz were relatively rough with attached matrix debris.

One possible explanation for these aforementioned differences between low (1 and 100 Hz) and high (900 Hz) frequencies is that there was lot more internal sliding and friction between fiber and matrix in the case of highest frequency, which was evident from the surface temperature measurements. This probably generated much higher temperature in the material especially in the sliding contact regions, such as in the fiber/matrix interface. This intense local increase in temperature possibly caused some interfacial reaction in the fiber/matrix interfacial region which resulted in the local interfacial bonding during the cycling.

CONCLUSION

Fatigue life of Nextel720^{TM}/alumina increased considerably at the highest loading frequency (900 Hz) while there was relatively less increase in the fatigue life with increasing cyclic frequency between 1 and 100 Hz at lower stress levels and practically none at higher stress levels (i.e. greater than 130 MPa). There was no difference in the failure and damage mechanisms with increase of frequency between 1 and 100 Hz. However, there was a strengthening of fiber/matrix interfacial bonding due to frictional heating at the highest frequency of 900 Hz. This interfacial strengthening phenomenon was most probably the cause of increase in the fatigue life/strength of the tested CMC system at the highest frequency of 900 Hz.

REFERENCES

[1]J. M. Staehler, and L. P. Zawada, The Performance of Four Ceramic-Matrix Composite Divergent Flaps Following Ground Testing on an F110 Turbofan Engine, *J. Am. Ceram. Soc.,* **83**, 1727-1738 (2000).
[2]R. A. Jurf, and S. C, Butner, Advances in Oxide-Oxide CMC, *Am. Soc. Mech. Engn., The 44th ASME Gas Turbine and Aeroengine Symposium,* June 7-10, 1999, Indianapolis, Indiana.
[3]P. Spriet, and G. Harbarou, Applications of CMCs to Turbojet Engines: Overview of the SEP Experience, *Key Engineering Materials,* **127-131**, 1267-76 (1997).
[4] S. Steel, L. P. Zawada, and S. Mall, Fatigue Behavior of Nextel 720/Alumina (N720/A) Composite at Room and Elevated temperatures, *Ceram Eng. Sci. Proc.* **22**, 695-702 (2001).

FRICTION AND WEAR BEHAVIOR OF AlBC COMPOSITES

Ellen Dubensky, Robert Newman, Aleksander J. Pyzik and Amy Wetzel
The Dow Chemical Company, New Products R&D
Midland, MI 48764

Abstract
Aluminum – Boron Carbide (AlBC) composites represent a family of light weight materials for which the composition and properties can be tailored to the requirements of specific applications. Some of these applications require the ability to operate at temperatures higher than those considered feasible for Al metal and alloys. For these applications AlBC composites have been developed which have high strength above the melting point of aluminum due to the strongly interconnected multiphase ceramic network [1]. The interconnected network of hard B_4C-based ceramic phases should make AlBC an attractive material for friction and wear applications. However, evaluating the friction & wear properties for AlBC composites is often difficult, because it requires fabrication of large, often complex full-size test parts. In this paper we investigate the capability of a small-scale test system to simulate the friction & wear conditions of a full size test system, specifically for automotive brakes.

Two types of AlBC composites were studied: GEN3 had higher toughness & lower hardness/stiffness; GEN4 had higher hardness/stiffness, but lower toughness [2]. The friction and fade behavior of the two compositions was evaluated versus cast iron, and showed that GEN4 had high and stable friction performance up to 750°C. In addition, the experiments provided valuable information regarding tribology and fracture behavior of AlBC materials under the aggressive thermal and mechanical stresses of the disc brake environment.

1.0 Introduction
1.1 AlBC Material Background

Aluminum Boron Carbide composites (AlBC) represent a family of light weight materials where multi-phase compositions are produced by reacting boron carbide (B_4C) ceramics with liquid aluminum [2-4]. These composites are characterized by very attractive room temperature properties, such as high hardness (1400-1600 kg/mm^2), high Young's modulus (320-360 GPa), high strength (500 MPa) and excellent wear resistance [5]. Many applications, such as automotive disc brake systems, that could benefit from light weight materials also require high temperature stability. It is well known [6] that boron carbide itself oxidizes readily above 600°C forming volatile B_2O_3. However, it has also been reported that the addition of Al and Si dopants improve the oxidation resistance of boron carbide, enabling the material to better retain strength at higher temperatures [7]. AlBC contains both aluminum and small amounts of silicon. In contrast, however, to hot pressed boron carbide, AlBC composite is a complex, multiphase material which contains the reaction products of B_4C and Al: AlB_2, $AlB_{24}C_4$ and $Al_{3-4}BC$. These phases, plus residual B_4C and Al determine the rate of surface oxidation and subsequent high temperature properties of AlBC composite. The change in the chemistry and microstructure, due to the oxidation, should have a direct effect on the surface region of an AlBC material, and as a result should play a major role in determining the friction/wear properties of an AlBC wear surface at typical operating temperatures of high performance automotive disc brake systems.

1.2 High Performance Brake Application

Ceramic-based materials are of interest for automotive high performance brake discs due to their light weight, high wear resistance and high temperature stability compared to conventional cast iron and steel. Carbon-SiC composites are reported to deliver outstanding braking performance including shorter stopping distance and little or no fade, or decrease in friction coefficient with increasing rotor temperature [8]. Temperatures at the brake disc surface can reach 800°C or higher under extreme braking conditions and candidate rotor materials must be able to maintain friction performance and dimensional stability under these conditions.

Automotive disc brake companies typically test new materials and designs using full-scale rotors on dynamometers. Standard tests such as SAE J2522 and ISO NWI 2560 are used to evaluate friction performance under a wide range of driving/stopping conditions. Full-scale dyno testing is preferred due to the influence of thermal mass and design features such as internal vent geometry on cooling and friction performance. However, for early-stage evaluation of new disc materials, such as ceramic matrix composites, the full-scale dyno tests are expensive and time consuming due to the difficulty in designing and fabricating full-size rotors. While material properties such as modulus, toughness and strength in the temperature range of interest can give some indication of a material's potential applicability, they do not adequately predict rotor material performance due to the complex and changing conditions at the friction interface during braking. A desirable material screening test would utilize relatively simple specimen geometries for the disc and pad materials, and would also simulate realistic speed and pressure conditions at the friction interface.

Several custom brake disc tests have been developed and reported for the purpose of material screening [9-11]. Most of these either do not simulate the true brake environment well, particularly the high temperature at the friction interface, or require relatively large and/or complex test part geometries. A more recently developed system by Kermc, et. al. at the Center for Tribology and Technical Diagnostics, University of Ljubljana, addresses these issues [12]. This system is able to simulate the temperature & friction conditions typically observed during brake testing on full size dynamometers, but utilizes a small size, simplified test geometry for friction test samples.

The purpose of this paper is to evaluate the capabilities of this small-scale test system in simulating the full-scale dynamometer test environment, summarize the evaluation of two different compositions of AlBC composite (GEN3 and GEN4) as potential disc brake friction surfaces and to understand the changes in (1) material properties (toughness, modulus, strength), (2) friction/wear properties, and (3) surface composition/chemistry under simulated disc brake operating conditions of high temperature & high friction.

2. Experimental Procedures
2.1 Raw materials and part preparation

A blended composition of ESK F1500 and F1200 grade powders was used in this work. The ESK F1500 grade powder had a 1.24 wt% oxygen and 20.75 wt% carbon content. The main impurities were 750 ppm of Fe, 460 ppm of Si, 210 ppm of Ti, 200 ppm of Ca and 90 ppm of Zn. The average particle size (d_{50}) was ~1 μm with a surface area of 13.6 m^2/g. The ESK F1200 grade powder contained 1.4 wt% oxygen and 21.21 wt% of carbon. The main impurities were 450 ppm of Ca, 270 ppm of Si, 165 ppm of Fe, 80 ppm of Mg and 65 ppm of Ti. The average particle size (d_{50}) was ~1.7 μm with a surface area of 6.9 m^2/g. Porous B_4C preforms made utilizing these two grades of ESK powders were infiltrated with 6061 grade aluminum alloy with the composition shown in Table I.

The AIBC GEN3 and GEN4 composites were made by slip casting using a 35 to 40 vol% slurry containing a blend of 50% ESK F1500 and 50% ESK F1200 powders. The ESK powders were washed in methanol and dried under nitrogen prior to use in slurry formulation, in order to remove boron oxide (B_2O_3) known to degrade dispersability of B_4C powders in water [13] (subsequent surface re-oxidation during slurry preparation and drying was relatively low).

Table I. Composition of 6061 Grade Aluminum Alloy

Component	Wt. %	Component	Wt. %	Component	Wt. %
Al	95.8 - 98.6	Mg	0.8 - 1.2	Si	0.4 - 0.8
Cr	0.04 - 0.35	Mn	Max 0.15	Ti	Max 0.15
Cu	0.15 - 0.4	Other, each	Max 0.05	Zn	Max 0.25
Fe	Max 0.7	Other, total	Max 0.15		

Once fully dispersed to the desired vol% loading, slurries were poured into 2 liter bottles containing B_4C milling media, and rolled for a minimum of 48 hours prior to use for casting parts. Prior to casting, slurries were filtered, then degassed under vacuum. After casting in a mold on a plaster block, porous B_4C parts (of varying sizes, from ~3 inches by 3 inches by 0.5 inch up to 13" diameter rings) were allowed to slowly dry over a period of 2-3 days. Drying behavior was controlled by using Plexiglas covers on the casting molds, and also enclosing the entire plaster casting block in a plastic bag when the ambient humidity was low. When adequately dried, parts were removed from the plaster block and placed in a nitrogen-purged drying oven at 45°C for a minimum of 24 hours.

After the parts were fully dried, they were heat treated above 1000°C in a furnace under an Ar/5% H_2 atmosphere with a residence time at heat treat temperature of 1-2 hours. After heat treatment, the infiltration furnace was loaded with the porous B_4C parts, with the appropriate amount of aluminum 6061 alloy ingot placed in contact with each individual part. The furnace was evacuated to 1 torr, backfilled with Ar gas, and then pumped to high vacuum (10^{-4} torr) for the aluminum infiltration furnace run. The residence time at infiltration temperature was typically 1 hour. After infiltration, the GEN3 composition was post heat-treated in an air furnace at 800°C for several hours to obtain the GEN4 composition.

2.2 Material Properties

The analyses to obtain composition and physical properties of the AIBC GEN3 and GEN4 compositions were reported previously [2]. Results of the analyses of these two compositions are shown in Table II.

2.3 Small-scale friction testing

The sample geometry for the small-scale disc is shown in Figure 1. The sample outer diameter was 60 mm, with inner diameter of 30 mm and thickness of 9mm. The sample had three circular notches in the inner diameter for locating on the drive shaft. Cast iron reference disc samples were machined from commercially available cast iron brake rotors.

The brake pad samples were 16mm X 8 mm with a 0.2mm thermocouple hole drilled partially through the center. The brake pad samples were fabricated from commercially available semi-metallic (Wilwood Polymatrix 7112E) brake pads. Two pad samples were used for each disc sample, as shown in Figure 2.

TABLE II. Summary of Properties for AlBC Compositions GEN3 and GEN4*.

Symbol	GEN3 Composition AlBC					
	Temperature [°C]	20°C	100°C	250°C	500°C	750°C
σ_{flex}	Flexural strength [MPa]	614 (±55)	567 (est.)	478 (±33)	336 (±26)	106 (±3)
E	Young Modulus [GPa]	215 (±15)	210 (est.)	201 (±11)	130 (±7)	106 (±4)
K_{Ic}	Fracture toughness, K_{Ic} [MPa/m²]	9.8 ±1.2			5.4 ± 0.02	1.5
ν	Poisson ratio, ν	0.26				
H	Hardness [Vickers, kg/mm²]	700 ± 90				
C_P	Specific heat capacity [J/g -K]	0.9 (est.)	1.095	1.36	1.65	1.77
λ	Thermal conductivity [W/m − K]	59.3	58.3	54.8	49.9	29.8

Symbol	GEN4 Composition AlBC					
	Temperature [°C]	20°C	100°C	250°C	500°C	750°C
σ_{flex}	Flexural strength [MPa]	472 (±30)	467 (est.)	457 (±44)	413 (±37)	313 (±34)
E	Young Modulus [GPa]	303 (±18)	301 (est.)	297 (±15)	281 (±9)	257 (±12)
K_{Ic}	Fracture toughness, K_{Ic} [MPa/m²]	5.4 ± 0.1			3.8 ± 0.1	2.9 ± 0.1
ν	Poisson ratio, ν	0.22				
H	Hardness [Vickers, kg/mm²]	1450 ± 100				
C_P	Specific heat capacity [J/g -K]	0.9 (est.)	1.13	1.38	1.69	1.94
λ	Thermal conductivity [W/m − K]	32.9	30.9	28.5	26.2	24.3

*Note that these two compositions of AlBC were previously designated as "B/C" and "C/D" in an earlier study [2].

2.4 Test Equipment

The small-scale friction test set-up at the University of Ljubljana was used for testing the AlBC and cast iron samples. Details of the test, including temperature measurement and dimensions of the pad and disc samples can be found in [12]. Figure 2 shows a disc sample mounted in the rotating spindle and the mating pad samples. The temperature close to the friction surface was measured using a thermocouple placed into a hole drilled into the pad samples, with approximately 0.5 mm of pad material remaining between the thermocouple and friction surface.

Figure 1. Small-scale friction test specimen

a   b

Figure 2. Small-scale friction test apparatus: (a) mounted disc sample after testing.
(b) mounted brake pads after testing

2.5 Test Procedures
Constant Load Test Procedure

For the constant load test, the brake pad samples were pressed at a constant load against the disc which was rotating at a constant velocity, for a period of 100 seconds. The pads were then pulled off the disc surface, the disc rotation stopped, and the disc temperature cooled to 50°C. The 100 second constant braking cycle was repeated 5 – 6 times for each combination of disc speed and pad pressure. Friction surface temperature and friction coefficient were recorded throughout the testing procedure. Results from this test were recorded as average friction coefficient from the last 4 – 5 braking cycles, since the first 1 or 2 braking cycles typically showed low friction coefficient as a result of only partial contact of the pad samples with the disc surface. After one or two braking cycles, the pad samples are worn enough to give full contact of the pads on the disc. An example of the results from six cycles of the constant load test for a cast iron reference sample is shown in Figure 3.

Procedure for Temperature Step Test

The temperature step-test allows for a more gradual increase in temperature at the friction interface, while still measuring friction coefficient vs. temperature. [14] For this test, the pad samples were pressed against the rotating disc sample at constant pressure and disc velocity, and the braking time was controlled by a preset temperature limit. Using this method, the temperature at the friction surface was gradually increased to 700 C in 100 increments, and the results were recorded as average friction coefficient for 10 cycles at each temperature increment.

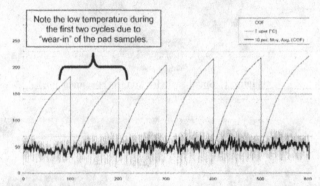

Figure 3. Pad temperature and coefficient of friction for cast iron as a function of test time. Test velocity was 6.0 m/s and pressure was 0.39 MPa.

3.0 Results and Discussion

3.1 Friction/fade

The friction coefficient results for cast iron and AIBC GEN3 using the constant load test procedure are shown in Figures 4 and 5 respectively. Both the cast iron and AIBC GEN3 show typical fade, or decrease in friction coefficient, as the severity of braking conditions, or temperature at the interface increases. The maximum temperature measured at the interface with cast iron was 800°C at the highest pressure and highest speed condition (1.17 MPa, 14.64 m/s).

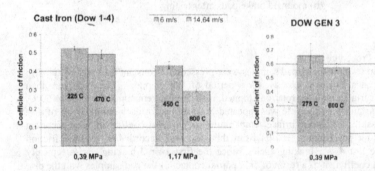

Figure 4. Results from constant load test for cast iron.

Figure 5. Results from constant load test for AIBC GEN3.

At this condition, the friction coefficient measured was 0.3. For AIBC GEN3, the highest temperature measured at the interface was 650°C, also at the high speed, high pressure condition. Under these conditions, the friction coefficient of GEN3 was the same as cast iron, or 0.3.

Compared to cast iron, the GEN3 AIBC showed significantly more degradation and wear under the high speed/pressure conditions. Above about 650°C, the residual aluminum in GEN3 began to melt and ooze from the disc surface, causing accelerated wear. Figure 6 describes in further detail the observations that were made for GEN3 samples under high temperature test conditions.

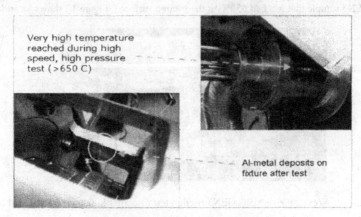

Figure 6. Observations for AIBC GEN3 friction test disc
under high temperature test conditions.

The friction coefficient results for AIBC GEN4 using the temperature step test procedure are shown in Figures 7 and 8 for two different pressure/speed conditions. In both samples, the friction coefficient remained stable, at about 0.6, up to 700°C. This is in contrast with the cast iron and AIBC GEN3 samples which both showed decreasing friction coefficient with increasing

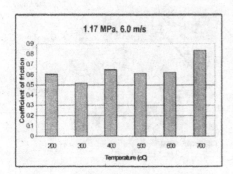

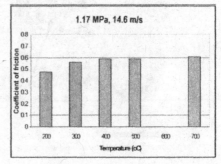

Figure 7. Results from step-test for AIBC
GEN4 at 1.17MPa, 6.0 m/s.

Figure 8. Results from step-test for AIBC
GEN4 at 1.17MPa, 14.6 m/s.

temperature, and it is consistent with what is reported in the literature for C-SiC [12]. The AlBC GEN4 samples fractured during the step-test, but were held together with a clamp in order to complete the test.

3.2 Analysis of friction surface/microstructure

Compared to cast iron, the AlBC GEN3 showed significantly more degradation and wear under high speed/pressure conditions. Figure 6 describes the observations that were made for AlBC GEN3 samples under high temperature test conditions, and Figure 9 shows the wear track in AlBC GEN3 sample that reached 650°C at the friction surface. Figure 10 shows 5x and 40x

Figure 9. Wear track in AlBC GEN3 friction test disc after high temperature load test.

micrographs of the polished AlBC GEN3 sample cross-section taken through the wear track, which shows the porosity which remained after melting and migration of the residual aluminum to the surface (see transfer layer in 40x image). Note the high degree of porosity, as shown in the micrographs. This highly porous microstructure explains the gouged appearance and significant wear of the AlBC GEN3 disc.

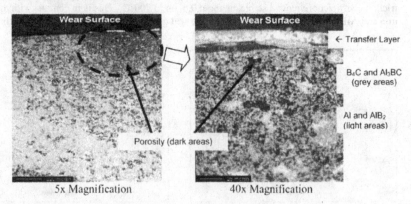

Figure 10. Electron micrographs of polished AlBC GEN3 sample cross-section taken through the wear track.

Figure 11 shows a micrograph of a polished cross-section through the wear track of an AlBC GEN4 sample. The figure shows a transfer layer of pad material (visible in the 40x and 100x images) that has been deposited on the wear surface. Underneath the pad transfer layer, the GEN4 material shows no significant surface degradation of the AlBC microstructure.

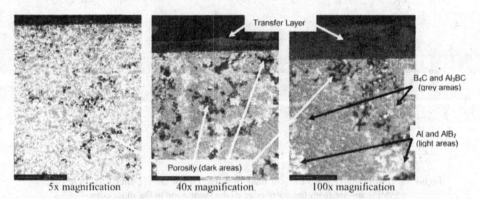

Figure 11. Electron micrographs at 5x, 40x and 100x magnification of polished cross-section through the wear track of AlBC GEN4. The darker grey regions at 100x are B_4C; lighter grey regions at 100x are Al_3BC.

Similar to the GEN3 test samples, the GEN4 material does exhibit porosity, but the porosity is much sparser, with the majority of the material (dense ceramic composite) remaining intact. The darker grey areas in the 100x micrograph are unreacted B_4C and the lighter grey areas surrounding the B_4C are the reaction phase Al_3BC. The grains that remain in the center of the porous region near the wear surface in the 100x micrograph are also unreacted B_4C.

3.3 Profilometry Analysis of AlBC GEN4 friction surface
Due to the extreme wear and gouging of the AlBC GEN3 sample, only the ALBC GEN4 wear surface was analyzed using scanning profilometry. The results for the AlBC GEN4 sample are shown in Figure 12. Over the ~8 mm wear area, three of the four scans from different locations show that the marks observed on the disc surface are generally material that has been added to the surface, standing a few hundred microns high. Few gouges reach all the way to the ALBC substrate or below. The fourth scan (lower right) does show scouring into the ALBC substrate to a depth of ~ 200 microns.

3.3 Backscatter Electron (BSE) Imaging of AlBC GEN4 friction surface
The backscatter electron images obtained from a ~500 micron wide segment of the wear track of the AlBC GEN4 disc sample are shown in Figure 13. Note the variations in image brightness in image 13a are due to differences in elemental composition of the surface, which we believe is primarily from surface deposition of different pad materials & metal (Fe, Al) oxides. The variation in topography shown in the secondary image is also likely due to surface deposition of pad material observed on the left hand side of the image.

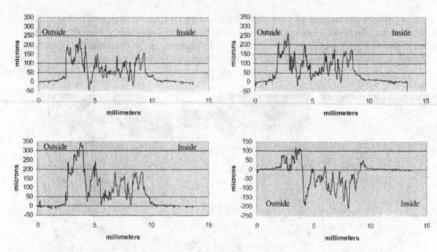

Figure 12. Profilometry scans of wear track of AlBC GEN4 sample after friction testing.
All scans are from the outer edge of the wear zone to the inner edge.

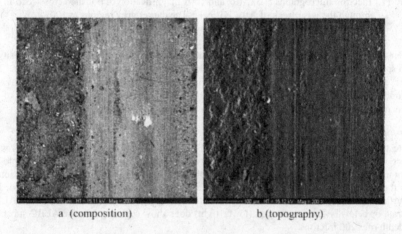

a (composition)   b (topography)

Figure 13. Primary (a) and Secondary (b) Backscatter Electron (BSE) Images of wear track
in AlBC GEN4 disc sample after friction/wear test.

Electron-Dispersive Scattering (EDS) spectra obtained from several regions across the wear
track indicated the presence of iron oxides, aluminum (metal and oxide) and zinc (metal and
oxide). A typical EDS spectrum from the AlBC GEN4 wear surface is shown in Figure 14.
Note that there is no obvious contrast in the image that defines the region containing oxidized
iron and aluminum.

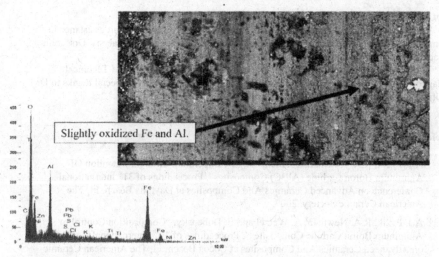

Figure 14.   Electron Dispersive Scattering (EDS) spectrum obtained from a region of a BSE image of the wear track in AlBC GEN4 disc sample after friction/wear test.

3.4 Conclusions

The friction/fade performance of AlBC GEN3 and GEN4 materials was assessed using the small-scale friction test system developed by Prof. Kalin at the University of Ljubljana, Slovenia, yielding the following conclusions:

1. Cast iron showed decreasing friction coefficient with increasing surface temperature, which is consistent with published results, and represents typical "fade".

2. AlBC GEN4 showed high and stable friction coefficient up to about 700°C with little or no visible wear. This is similar to results published for C-SiC, which is currently being used in high-performance brake rotors. The GEN4 samples fractured when the disc temperature exceeded about 350°C, so a clamping device was used in order to complete testing.

3. AlBC GEN3 showed a decrease in friction coefficient with increasing surface temperature, or typical fade. GEN3 showed rapid wear above the melting point of aluminum (about 600°C), as molten aluminum oozed from the disc surface.

The results from testing of GEN3 and GEN4 materials indicated that neither was suitable for the brake application. The testing did show that the small-scale friction test system developed by Prof. Kalin at the University of Ljubljana is capable of simulating the friction/wear environment of a full-size dynamometer test system.

Acknowledgements
The authors express their sincere thanks to the following people for their assistance in completing this study: Drs. Cliff Todd for profilometry, SEM and EDS analysis; Drs. John Blackson, Kalyan Sehanobish and Tony Samurkas for helpful discussions.
The assistance and expertise of the staff at the Center for Tribology and Technical Diagnostics, University of Ljubljana is gratefully acknowledged, with a special thanks to Dr. Mitjan Kalin for coordinating the friction tests.

References

1. A.J. Pyzik, R.A. Newman, S. Allen, "High Temperature Strength Retention Of Aluminum Boron Carbide (AlBC) Composites", Proceedings of 31st International Conference on Advanced Ceramics And Composites at Daytona Beach, Fl, The American Ceramic Society, 2007.

2. A.J. Pyzik, R.A. Newman, A. Wetzel and E. Dubensky, "Composition Control in Aluminum Boron Carbide Composites", Proceedings of 30th International Conference on Advanced Ceramics And Composites at Cocoa Beach, Fl, The American Ceramic Society, 2006.

3. D.C. Halverson, A.J. Pyzik, I. A. Aksay, W.E. Snowden, "Processing of Boron Carbide – Aluminum Composites", J. American Ceramic Society, 72, (5), 775, 1989.

4. I. A. Aksay, D.M. Dabbs, J.T. Staley, M. Sarikaya, "Bioinspired Processing of Ceramic Metal Composites", Third Euro-Ceramics, ed. P. Duran and J. F. Fernandez, vol. 1, pp.405-418, 1993.

5. A. J. Pyzik, D. R. Beaman, "Al-B-C Phase Development and Effects on Mechanical Properties of B4C/Al Derived Composites", Journal American Ceramic Society, 78, (2), pp.305-12, 1995.

6. G.N. Makarenko, "Borides of the IVb Group, in Boron and Refractory Borides", edited by V.I. Matkovich, Springer-Verlag Berlin Heidelberg New York, 1977, pp. 310-330.

7. Yu. G. Gogotsi, V.P. Yaroshenko, F. Porz, "Oxidation Resistance of Boron Carbide-Based Ceramics", Journal of Materials Science Letters, 11, 308-10, 1992.

8. Krupka, M. and Kienzle, A., Fiber reinforced ceramic composite for brake discs. Proceedings of the 18th Annual Brake Colloquium and Engineering Display. SAE Paper No. 2000-01-2761, SAE, Warrendale, 2000, p 358.

9. P.J. Blau. In: Compositions, functions, and testing of friction brake materials and their additives, Oak Ridge National Laboratory, Oak Ridge, Tennessee (2001), p. 24 [Report ORNL/TM /64] .

10. P. Blau and B. Jolly, "Wear of truck brake lining materials using three different test methods", Wear, Volume 259, Issues 7 - 12, July-August 2005, Pages 1022-1030.

11. P. Sanders and R. Basch, "A reduced-scale brake dynamometer for friction characterization", Tribology International, Volume 34, Issue 9, September 2001, pp 606-615.

12. Kermc, Kalin, Vizintin, "Development and use of an apparatus for Tribological evaluation of ceramic-based brake materials", Wear, Volume 259 (2005), 1079 – 1087.

13. J.D. Stachiw, A.J. Pyzik, D. Carroll, A. Prunier, T. Allen; Boron Carbide Aluminum Cermets for External Pressure Housing Application, Technical Report 1574, Naval Command, Control and Ocean Surveillance Center, San Diego, 1992.

14. M. Kalin, to be published.

CREEP OF SILICON NITRIDE OBSERVED IN SITU WITH NEUTRON DIFFRACTION

G. A. Swift[*,**]
Department of Materials Science, California Institute of Technology
Pasadena, CA 91125

ABSTRACT

The microstructure of GS-44 silicon nitride during high temperature creep was observed *in situ* with neutron diffraction. Using the SMARTS diffractometer at the Los Alamos Neutron Science Center, constant stress tensile creep studies were performed on GS-44 at 1200°C. Observed diffraction spectra from the longitudinal detector banks demonstrated a unique microstructural effect in the lattice parameter for those grains aligned with the applied stress. The *a* and *c* lattice parameters fork, with the *a* parameter increasing and the *c* parameter decreasing. In accord with mathematical models, the forking seeks to preserve a constant volume of the unit cell averaged over the observed grains. The lattice parameter forking is more pronounced at higher applied stress. The creep exponent was 3.18, higher than literature values due to the experimental conditions. A creep model for *in situ*-reinforced Si_3N_4 was able to match the steady state creep rate data, but required an empirical relation of the applied stress to do so.

INTRODUCTION

Silicon nitride (Si_3N_4), when densified properly, can achieve an *in-situ* reinforced (ISR) microstructure, as with GS-44 (Honeywell Ceramic Components, Torrance, CA). This reinforcement is accorded to the acicular grain structure generated by the processing conditions, *e.g.*, hot pressing or gas-pressure sintering. Such Si_3N_4 typically exhibits improved mechanical properties as a result of this microstructure, and thus is referred to as *in situ*-reinforced, or ISR [1]. Improvements to creep resistance are believed to be caused by grain rearrangement and interlocking of the elongated grains in the microstructure or by a solution-precipitation process; regardless of the creep mechanism, the creep experienced depends heavily on the grain boundary phase behavior [2]. Present knowledge is unclear as to the exact mechanism that dominates the creep process, primarily due to a lack of *in-situ* test methods. The steady state creep rate has traditionally been modeled using the Norton equation:

$$\dot{\varepsilon} = \frac{ADGb}{kT}\left(\frac{b}{d}\right)^{p}\left(\frac{\sigma}{G}\right)^{n} \tag{1}$$

in which: σ is the applied stress, D is the diffusion coefficient, G is the shear modulus, b is the Burger's vector, k is the Boltzmann constant, T is the absolute temperature, d is the average grain size, p is the inverse grain exponent, n is the stress exponent, and A is a constant. Luecke and Wiederhorn developed a model for tensile creep of ISR Si_3N_4 based on observations of the creep behavior, including the formation of cavities in the grain boundary phase, the inactivity of dislocations, and the curvature of plotted tensile creep data [3]. Their creep equation is shown in Equation (2),

$$\dot{\varepsilon} = A\sigma\exp\left(\frac{-\Delta H}{RT}\right)\frac{f^{3}}{\left(1-f\right)^{2}}\exp(\alpha\sigma) \tag{2}$$

in which f is the volume fraction of second phase, α is a constant incorporating the critical stress to nucleate a cavity, ΔH is activation energy, R is the gas constant, and the other terms as for the Norton equation. For convenience, since the present measurements were from the same material and at the same temperature, the various constants were combined into a single constant:

$$\dot{\varepsilon} = A'\sigma\exp(\alpha\sigma) \tag{2a}$$

131

Luecke and Wiederhorn indicated that the Si_3N_4 grains remain elastic while creep proceeds via the grain boundary phase. Thus, diffraction should be able to observe the effects in the grains since elastic strains (but not plastic strains) can be measured.

The ambiguity in determining the creep mechanism lies in the *ex-situ* nature of the experiments, as regards examination of microstructure. Until recently, there has been a lack of equipment capable of probing the microstructure of Si_3N_4 under creep conditions, so the microstructure has only been examined after creep has occurred. SMARTS (Spectrometer for Materials at Temperature and Stress), through integration of a vacuum furnace (MRF Inc., Suncook, NH) and custom load frame (Instron Corp., Canton, MA), is able to probe the microstructure of bulk samples with time-of-flight neutron diffraction (ND) while applying tensile stress at high temperature [4]. SMARTS was applied in this manner to study the microstructure of GS-44 under tensile creep at 1200°C. Results are presented for longitudinal strains for several creep tests.

Prior work [5] demonstrated the utility of the SMARTS system for analyzing the effects of high-temperature tensile stress application on the microstructure of AS800 silicon nitride. While creep had been expected at the test conditions, only elastic strains were observed. While this allowed computation of the high temperature stiffness tensor for AS800, the experiment was unsuccessful in observing creep. Atmospheric oxygen reduces the viscosity of the grain boundary phase in $ISR-Si_3N_4$, thus creep was prevented due to the vacuum furnace used to achieve high temperature. A grade of silicon nitride that creeps at lower temperature was required for the experiment to proceed; GS-44 was selected for this purpose [6].

This paper presents the results from neutron diffraction showing the microstructural effects of GS-44 Si_3N_4 due to tensile creep at high temperature in vacuum.

EXPERIMENTAL PROCEDURE

The specimens were standard pin-loaded dog-bone samples of GS-44 ISR Si_3N_4 with a gage section of 19.2 mm^2. Samples were subjected to tensile stresses of 100, 125, 150, and 175 MPa in the SMARTS furnace after reaching 1200°C. Samples were mounted at a 45° angle relative to the incident neutron beam, to allow the collection of longitudinal diffraction data [5]. After loading a sample into the tensile grips, the furnace was evacuated to a pressure of <0.013Pa for the duration of the experiment, and a tensile stress of 25-30MPa was applied to prevent sagging. Heating was gradual, with pauses approximately every 200°C to allow for thermal equilibration. Upon reaching 1200°C, a reference diffraction pattern was collected (for calculating lattice strain) and the stress was stepped up to the creep stress incrementally to avoid fracture. Once the creep stress was reached, it was held constant for up to 40 hours with an unload/reload cycle during creep (two unload/reload cycles were performed for the 175MPa sample).

Diffraction spectra indicated only $\alpha-Si_3N_4$, with no reflections apparent from the grain boundary phase, an H-phase apatite glass [1], indicating no crystallization of said phase. A SiC-blade high-temperature extensometer monitored the bulk strain of the sample during the entire experiment.

Note that the maximum use temperature of GS-44 per the manufacturer is 1100°C. However, as this experiment was to take place in vacuum, there would be a greater viscosity of the grain boundary phase compared to a test in air, due to a lack of available oxygen [7]. Thus, a higher temperature was chosen to facilitate the onset of creep. Diffraction spectra were obtained every 20 minutes during creep. This acquisition time was the minimum necessary for high-quality patterns, thus each spectra sums the behavior over the 20 minute time span.

RESULTS AND DISCUSSION

Figure 1 shows the creep strains as measured with the SiC extensometer prior to any unload/reload cycle. Creep is evident, though of a much lesser magnitude than seen in literature for this grade of Si_3N_4. Figure 2 shows the log-log plot of the creep strain rates (obtained from Figure 1) versus the applied tensile stress. The creep exponent is 3.18, greater than the 2.24 value for GS-44 measured by others [6]. Again the inhibition of creep due to the present experiment taking place in vacuum rather than in air prevented oxygen from reducing the grain boundary phase viscosity.

Figures 3-6 show the a and c lattice strains for GS-44 during creep at 1200°C. The gap in each plot during the creep test is from the unload/reload cycle that was performed. A data acquisition error for the 150MPa creep test resulted in the large gap prior to the unload/reload cycle. Each plot shows an average lattice parameter d, which remained mostly constant for the duration of the measurements. Evident in Figure 3 for 100MPa is a forking of the two lattice parameters. This forking only increases in separation with increasing stress as seen in Figures 4-6. While the a parameter shows an increase in strain, the c parameter shows a decrease. After some time under constant stress, the magnitude of the separation stabilizes and no longer increases, even though creep is still occurring per the extensometer; the forking remains after unloading/creep recovery as seen for all four applied stresses. Recall that only longitudinal lattice strains, for grains aligned with tensile direction, are shown here, thus the forking is not a Poisson effect.

The forking was found to be a volume conservation mechanism. The lattice parameter values were used to calculate the unit cell volume. These data, along with the c/a ratio, for the 175MPa creep sample are shown in Figure 7. Note that after the initial loading to the creep stress (the portion of the plot shown as negative time), the volume decreased slightly and reached a steady state value. After unload/reload, this steady state volume was reacquired. A second unload/reload cycle was performed with similar results. The unload/reload cycles for each sample indicated that 75MPa was the approximate critical stress, as neither creep nor creep recovery occurred at that stress. This was used to compare the present results with the Luecke-Wiederhorn model of Equation (2a). The model did not match the data, however, until an empirical relation was determined for the applied stress. This relation is given as Equation (3) and was used in place of the pre-exponential constant A' of their formulation. First, the constant was determined which allowed the model to match the 100MPa data. A linear relation to this value was then determined to allow the model to approximate the data for the other stresses.

$$A'(\sigma) = 0.08 * \left(\frac{\sigma}{25} - 3 \right) \tag{3}$$

The fit of the creep rate data to the Luecke and Wiederhorn model using the values of Equation (3) for each stress is shown in the semi-log plot of Figure 8. Clearly, the agreement between the adjusted model and the data is a good one. Note that while in this simplified analysis an additional stress dependence was added to the pre-exponential constant term, the precise nature of the inaccurate fit to the data without using this term remains to be determined. This might take the form of an additional stress factor in the overall equation of Luecke and Wiederhorn (Equation (2)).

CONCLUSIONS

GS-44 was creep tested at 1200°C in vacuum with neutron diffraction measuring the microstructural effects *in situ* for the first time. The a and c lattice parameters are seen to fork with greater magnitude at greater stress in such a way as to obtain and maintain a constant unit cell volume during steady state creep. The creep exponent was determined from extensometer data to be 3.18, higher than prior reported values due to the vacuum furnace used in the present study.

ACKNOWLEDGMENTS

This research was supported by NASA Glenn Research Center through grant number NAG3-2686. This work has benefited from the use of the Lujan Center at the Los Alamos Neutron Science

Center, funded by the DOE Office of Basic Energy Sciences and Los Alamos National Laboratory funded by the Department of Energy under contract W-7405-ENG-36.

FOOTNOTES

* Present address: EaglePicher Technologies, Defense and Space Power Division, Joplin, MO 64801.
** Member, The American Ceramic Society

REFERENCES

[1] C.J. Gasdaska, "Tensile Creep in an *In-Situ* Reinforced Silicon-Nitride", *Journal of the American Ceramic Society*, **77**(9), 2408-2418 (1994)

[2] Q. Wei, et al., "Microstructural Changes Due to Heat-Treatment of Annealing and Their Effect on the Creep Behavior of Self-Reinforced Silicon Nitride Ceramics", *Mat. Sci. and Eng.*, **A299**, 141-151 (2001)

[3] W.E. Luecke and S.M. Wiederhorn, "A New Model for Tensile Creep of Silicon Nitride", *Journal of the American Ceramic Society*, **82**(10), 2769-2778 (1999)

[4] M.A.M. Bourke, et al., "SMARTS - A Spectrometer for Strain Measurement in Engineering Materials", *Applied Physics A-Materials Science & Processing*, **74**, S1707-S1709 (2002)

[5] G.A. Swift, et al., "High-Temperature Elastic Properties of *In Situ*-Reinforced Si_3N_4", *Applied Physics Letters*, **82**(7), 1039-1041 (2003)

[6] Q. Wei, et al., "High Temperature Uniaxial Creep Behavior of a Sintered *In Situ* Reinforced Silicon Nitride Ceramics", *Ceramic Engineering & Science Proceedings*, ed. E. Üstündag and G. Fischman, **20**(3), 1999, The American Ceramic Society: Westerville, OH, 463-470

[7] G. Ziegler, "Thermo-Mechanical Properties of Silicon Nitride and Their Dependence on Microstructure", *Mat. Sci. Forum*, **47**, 162-203 (1989)

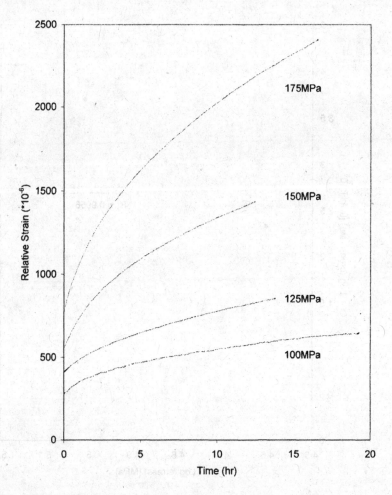

Figure 1. Extensometer strains for constant stress creep tests on GS-44.

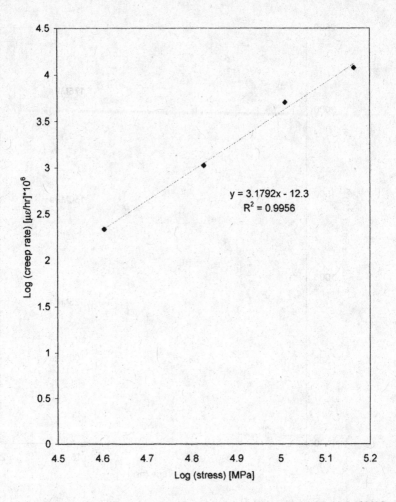

Figure 2. Log-log plot of steady state creep strain rate versus tensile stress for GS-44 at 1200°C.

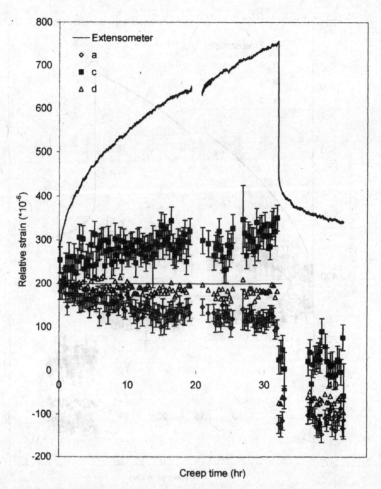

Figure 3. GS44 creep at 1200°C with 100MPa applied stress. Longitudinal diffraction strains are shown, relative to data from 25MPa, 1200°C. Horizontal line indicates the initial *d*-strain at 100MPa, 1200°C.

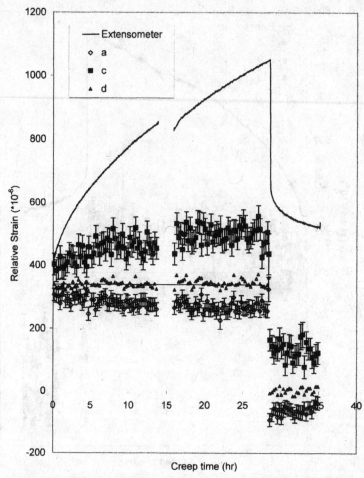

Figure 4. GS44 creep at 1200°C with 125MPa applied stress. Longitudinal diffraction strains are shown, relative to data from 25MPa, 1200°C. Horizontal line indicates the initial *d*-strain at 125MPa, 1200°C.

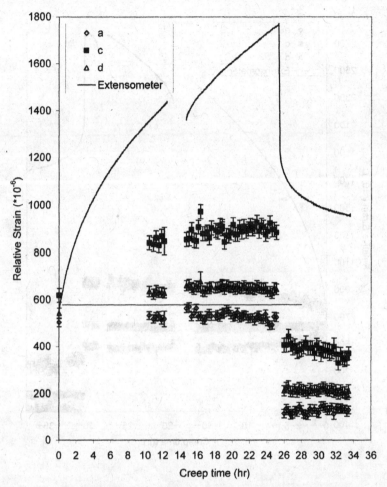

Figure 5. GS44 creep at 1200°C with 150MPa applied stress. Longitudinal diffraction strains are shown, relative to data from 25MPa, 1200°C. Horizontal line indicates the initial d-strain at 150MPa. 1200°C for the 175MPa test sample (since the first 150MPa data was lost for this sample).

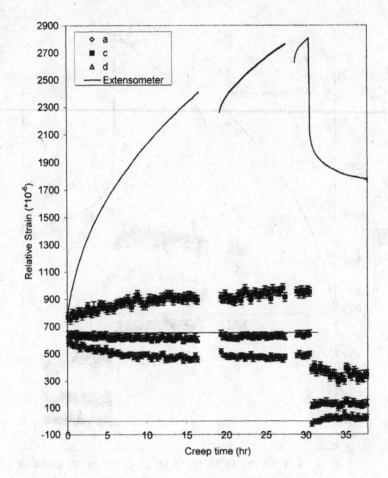

Figure 6. GS44 creep at 1200°C with 175MPa applied stress. Longitudinal diffraction strains are shown, relative to data from 25MPa, 1200°C. Horizontal line indicates the initial *d*-strain at 175MPa, 1200°C.

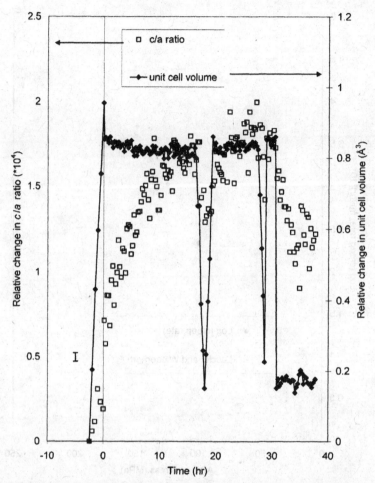

Figure 7. Relative change of *c/a* ratio and unit cell volume (relative to 1200°C, 25MPa diffraction data) for 175MPa creep of GS44. Data are from longitudinal 175MPa creep at 1200°C; a typical error bar is shown at left (± 5%), based on error from strains as seen in Figure .

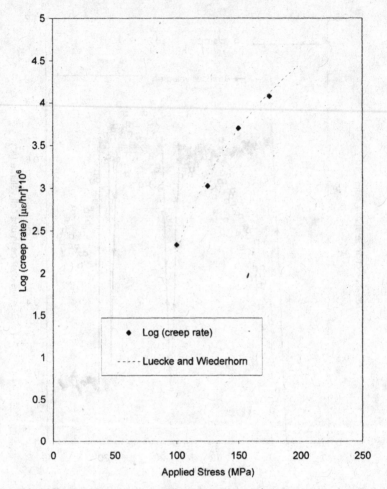

Figure 8. Semi-log plot of creep rate versus applied stress, with data prediction as per Equations (2) and (3). The predicted values approximate the data well (98.9% match).

HYDROTHERMAL OXIDATION OF SILICON CARBIDE AND ITS BEARING ON WET WEAR MECHANISMS

K.G. Nickel[1], V. Presser[1], O. Krummhauer[2], A. Kailer[2], R. Wirth[3]

[1]University Tübingen, Institute for Geosciences, Applied Mineralogy, Tübingen, Germany

[2]Fraunhofer-Institute for mechanics of materials, (IWM), Freiburg, Germany

[3]Geoforschungszentrum, Potsdam, Germany

ABSTRACT

Static hydrothermal oxidation experiments in a diamond anvil apparatus using wafer-quality single-crystal silicon carbide and observations from transmission electron microscopy from tribological tests on sintered silicon carbide are combined to gain insight into the wear process in water. The evidence points to an active type of oxidation under hydrothermal conditions, i.e. there are no primary condensed reaction products. The formed tribolayer is topographically highly variable and consists of disassembled silicon carbide (µm to nm-scaled), which contains secondary, precipitated silica in cracks and voids or as a glue in areas between original asperities. Hydrothermal pressures of several hundred MPa seem feasible and are seen as responsible both for fatigue and surface roughening as well as the formation of the smooth main wear track, which reduces friction favourably.

INTRODUCTION

Silicon carbide (SiC) is a material well known for excellent wear resistance in pumps and other devices, which are running in water or wet conditions. The sliding wear of SiC in vacuum or air[1, 2], oil[3, 4] and water [5-8] has been the subject of investigations over the last years. Such investigations usually report the development of surface topographies, friction coefficients and wear volumes after fixed times or sliding path lengths for a given loading and sliding speed.

It would be highly desirable to be able to predict wear rates from the loading data. There is however conflicting evidence from different views of looking at wear conditions. Following Quinn [9, 10] for metals and later Xu´s [11] adaptation for ceramics mainly the temperature arising from mechanical contact was taken into account for oxidation reactions with water to form a silica film. During ongoing mechanical contact latter is believed to partly delaminate and dissolved into the water to some extent.

The reactions discussed by Xu [11] for Silicon nitride were

$$Si_3N_4 + 6 H_2O \Leftrightarrow 3 SiO_2 + 4 NH_3 \qquad \text{(Eq.1)}$$

$$SiO_2 + 2 H_2O \Leftrightarrow Si(OH)_4 \quad \text{(Eq. 2)}$$

but the conditions of solubility were not discussed. In wet environments usually reaction (1) would cause a rise of the pH-value, because NH_4OH is formed from NH_3. A pH ≥ 10 is indeed required to raise the low solubility of silica from that of $Si(OH)_4$ with of about 10^{-3} m to levels three orders of magnitude higher[12], which are realised by other species like $H_2SiO_4^{2-}$ [13].

However, for silicon carbide the according reaction is

$$SiC + n\,H_2O \Leftrightarrow SiO_2 + CO_{n-2} + n\,H_2 \qquad\qquad (Eq.\ 3)$$

and could induce a slight decrease in pH instead of a rise. In acid environments silica stays at a very low solubility [12]. Furthermore, the kinetics of solution for silica, even in the amorphous state, are very slow and lie in the range of nm/a for pure natural silica (obsidian) below 100°C at pH 7 [14].

Nonetheless in a number of studies on water lubricated wear of SiC the loss of silica was reported and attributed at least partially to the solution of it in the water [5, 7, 15]. The solution to this problem could be hydrothermal conditions. At T > 100°C increased pressure is required to keep the water in a fluid state. From geoscientific evidence and experiments it is well known that the solubility and solution kinetics of silica rises with increasing pressure quite strongly [13, 16, 17]. Amorphous silica has here also a higher solubility relative to crystalline species [18].

Therefore the hydrothermal oxidation of silicon carbide could be a key to the kinetics and mechanisms of wear, if such conditions were present during the process. Earlier hydrothermal powder studies were interpreted to involve the formation of a silica film around SiC grains [19, 20]. However, later studies on powders, crystals and fibres showed an active oxidation mode, in which the reaction products were either completely dissolved or showed residual carbon or silica, which was re-deposited as a process of solution and precipitation [21-26]. These experiments were done at temperatures up to 800°C and several hundred MPa water pressure. Recently a study on CVD-SiC at 500°C was published [27], which gave indications for an active process, which however appeared to be slower than the data at lower temperatures of Kim et al. [28] suggested. The latter study also gave evidence for a still higher rate using SSiC instead of CVD-SiC. Therefore presently it is not clear, which process is dominating at which temperature and if it is an impurity related or intrinsic process.

Consequently Kitaoka performed wear tests under hydrothermal conditions [29, 30] but restricted the conditions to temperatures up to 300°C. He found active oxidation as a predominant mechanism.

Two of the most important key issues on the quest for a quantitative theory to predict wear rates of silicon carbide are the estimation of the true conditions (pressure, temperature, pH...) under ordinary wear conditions and the intrinsic behaviour of silicon carbide at severe hydrothermal conditions at higher temperatures. The study of the latter under static conditions is necessary to be able to build a wear theory for dynamic systems. Such questions are addressed in this paper.

THE CONDITIONS OF WEAR

A pure SiC – H₂O system, which is perfectly hydrodynamically lubricated at low temperatures will not show any wear worthwhile to be modelled. We are therefore talking about situations, which either reflect a "running in" of a system or a case, in which direct material contact is at least part of the process. When evaluating such wear conditions the conditions chosen for an experiment – i.e. experimentally boundary conditions which all serious reports quote – are certainly not those experienced at the material's outer surface. The reason for this is that in real surfaces consist of asperities. During wear those asperities are the only areas in direct mechanical contact and as such they collide and scratch each other.

There are two main pathways which a simulation of such conditions can follow. The first one is to come from the static point-of-view and to estimate the pressure from the theory of Hertzian contact pressures using simplified geometries for measured roughness profiles. The temperatures are then estimated from the boundary conditions such as shape, velocity and thermal diffusivity [31] [32] [33]. For a set-up of our experiments described below this results in a statistical distribution of P-T-conditions, which are often low in temperatures and very high in pressure with some peak values with very high temperature and extreme pressures (Fig. 1).

For a water lubricated system we may however estimate the conditions quite differently. Considering that temperature rises because mechanical energy is transformed into thermal energy and assuming that water can be trapped in a confined space (between asperities, in superficial pores and/or cracks) the pressure would rise only due to the isochoric heating process. These pressures are in the order of several hundred MPa and thus still very high but much lower then those running into tens of GPa from Hertzian theory (Fig. 1).

Hydrothermal oxidation under those conditions and temperatures of 500°C using a diamond anvil cell (DAC) has not been investigated before and is described and discussed below. Subsequently we will discuss the bearing of it on the water lubricated wear, which is tested alongside.

EXPERIMENTAL

Undoped 6H-SiC single crystal plates (SiCrystal AG, Erlangen, Germany) ground to a thickness of 100 μm and polished with 1 μm diamond paste were cut into small cuboids of approximately $240 \cdot 240 \cdot 100$ μm^3 to be used in hydrothermal runs. Each sample was carefully cleaned in tridestilled water and acetone for 30 minutes using an ultrasonic bath to remove remnants of the polishing paste.

For tribological tests EKasic ® F from ESK Ceramics GmbH & Co. KG (Kempten, Germany) was used. This material has fine grain size of about 5μm. It consists of app. 80 wt% 6H-SiC, 20 wt% 4H-SiC + 15R-SiC and additions of boron carbide as sintering aid. The sliding surfaces were lapped to values of R_z 2.79 μm and R_a 0.79 μm determined by a Hommel T 8000s profilometre (Hommel GmbH, Köln, Germany).

Hydrothermal experiments were performed with a hydrothermal diamond anvil apparatus (HDAC) after Basset et al. [34], which was modified and is described elsewhere [35]. Pure tridestilled water was used as a medium and the pressure was calculated with the aid of the equations of state (EOS) of pure water substance.[36, 37]. Samples were heated with 20 K·min^{-1} up to 500°C. This temperature, which corresponds to an isochoric pressure of 500 – 770 MPa, was maintained ± 1 K for 5 h before cooling to room temperature with the same gradient.

Tribological tests were conducted in a sliding ring tribometre described elsewhere [35] with deionized water as lubricant. During the experiment, the lubricant was applied with a pressure of 0.1 MPa between the friction couple.

Test runs in static and dynamic mode were executed at 0.5 MPa, external cooling and a sliding velocity of 6 m·s^{-1}. In intervals of ~ 2000 m sliding distance the sample surfaces were measured with the stylus profilometre and numerically evaluated. Experiments under dynamic conditions were carried out in absence of external cooling and with a sliding velocity of 6 m·s^{-1}. The abort criterion in case of scuffing was a measured torque of 2 N·m.

Sampling of an EKasic ® F sample for TEM analysis was accomplished by using focused ion beam (FIB) milling. [38] FIB preparation was conducted under ultra-high vacuum conditions in an oil-free vacuum system using a FEI FIB200 instrument (FEI Company, Tokyo, Japan) at the GeoForschungsZentrum Potsdam. As a result TEM-ready foils of approximately $20 \cdot 10 \cdot 0.15$ μm were obtained. The TEM foil was protected from abrasion by the Ga-ion beam by a 1 lm thick Pt layer deposited by using a high-purity organic Pt gas ($C_9H_{16}Pt$, 99.9%), which decomposes in the Ga-ion beam. The FIB-cut foils then were placed on a highly perforated carbon grid on a copper mesh. TEM was carried out in a Philips CM200 instrument operated at 200 kV and equipped with a LaB_6 electron source. Electron energy-loss spectroscopy (EELS) spectra were acquired with a Gatan imaging filter (GIF).

RESULTS

The hydrothermal treatment was monitored *in-situ* and showed an optical darkening of the transparent single crystals inside the diamond anvil cell's sample chamber, indicating increased roughness. A higher dehomogenization temperature during cooling was evidence for the separation of a carbonaceous species dissolved under hydrothermal conditions. Electron microscopy revealed the formation of pits and Raman spectroscopy indicated that no solid carbon layer was formed during the process. Silica was solely present as precipitates. The details will be described elsewhere [35].

The findings of Barringer et al. [27] of an active oxidation of CVD SiC at grain boundaries are hence confirmed for wafer quality single crystal SiC. It is an intrinsic process at temperatures up to 500°C.

From the tribotests without external cooling we chose samples, which were clearly affected as evidenced by a smooth tribological track surrounded by areas with clear evidence for roughening (Fig.2). To elucidate the process we took samples with the aid of a FIB and analysed the obtained slices with TEM methods.

Fig. 3 shows electron microphotographs of tribologically treated EKasic (R) F depicting the superficial tribolayer formed on smoothened areas (Fig. 3a – c) and on rough parts of the wear track (Fig. 3d – f).

Comparing the SEM and TEM results from smoothened and rough areas revealed differences in the tribolayer's thickness: on smoothened yet elevated areas the tribolayer was ~ 100 nm thick, while on rough areas a total thickness of ~ 500 nm could be observed. Deeper "valleys" between elevated parts, as seen in the right part of Fig. 3e, are filled with larger SiC particles forming a porous tribolayer with up to 1.5 μm total thickness. This is quite different to the nanoscale SiC particles, out of which the tribolayer in smoothened areas consists (Fig. 3c). In latter case SiC particles within the tribolayer have average diameters of 5 – 25 nm.

Fig. 4 depicts element mappings (C and O) of a smoothened (Fig. 4a) and rough (Fig. 4b) area. Oxygen-rich areas correspond with carbon-poor regions. These oxygen-rich areas act as a glue surrounding and keeping together rounded SiC wear particles. We also found oxygen-enriched areas along cracks formed during tribological exposure within the ceramic substrate.

While the tribolayer is limited to the very superficial area (= area of direct mechanical and chemical contact), the ceramic body itself shows mechanical damage and crack formation ranging much deeper into the substrate. In case of smoothened areas a 100 nm thick tribolayer is opposed by layer of visible mechanical damage (cracks) with a total thickness of 1 – 1.5 μm. Those cracks – as long as they are connected to the surface - can easily act as gateways for water to be transported deep into the ceramic body.

DISCUSSION

The overall process can now be viewed to involve the following steps (Fig. 5): Asperities of the material approach each other and collide. This is seen not just as a Hertzian pressing but as an impact event. The result is more then the generation of dislocations, because it shatters the structure of the material to disassemble the original carbide grains into smaller grains containing now cracks and cavities. Naturally there will also be erosional wear, because parts will spall and may be caught in the regions next to the asperities or carried away by the water. The rise in temperature due to the first impact may not be huge. When the asperities travel on and separate they regain the contact with water, which is now able to be sucked into the newly formed cracks and opened pores.

When the asperities meet again, we will have several changes. Further collisions will not only cause an intensified disassembling, which eventually produces a nano-scaled microstructure with very fine grains and domains and create even amorphous regions. It will also squeeze parts of the system to close former cracks. In this way these regions may become isolated from the rest. Naturally also the opposite will happen, where such regions become accessible by newly formed cracks. The energy of repeated contacts will tend to rise the temperature to a dynamic equilibrium value, controlled by the thermal diffusivity of the so formed tribolayer. Note that this must not be identical to that of the original material, because it contains voids and cracks and other phases.

Once isolated water bearing parts of the system are heated up we establish hydrothermal conditions with considerable pressures. Under these conditions we will have oxidation reactions combining reactions (2) and (3):

$$SiC + (n+2)\, H_2O \Leftrightarrow Si(OH)_4 + CO_{n-2} + n\, H_2 \qquad\qquad (Eq.\ 4),$$

from which silica and silicon oxycarbides could precipitate on cooling or depressurizing. Another form of oxygen storage could be the solution of it into amorphous carbide. .

The dynamics of the situation allow now several processes, which may become part of the wear process. One is temperature fluctuation. Temporarily or permanently falling temperatures should be accompanied by precipitation of silica and the evolvement of gas bubbles as a result of fluid dehomogenization. Temperature fluctuations would also cause pressure fluctuations and hereby induce fatigue phenomena in the already disrupted tribological layer. The level of pressures is in a range of up to several hundred MPa, which is high enough to cause cracking in tension. This should enhance the erosion and may be the reason for grain-pullout during wear. Likewise the formation of new cracks, along which a fluid could travel, is a reason for sudden pressure drops, which would be accompanied by both precipitation events and gas release. Precipitation of silica will be mainly in the amorphous state and could act as a good glue for the eroded particles, in particular those, which are too large to escape through the slit between the rubbing parts.

The texture of the FIB sections shown in Fig. 3 and 4 agree with those assumptions about the wear mechanisms. The oxygen rich parts occur along cracks and in isolated gussets in the tribolayer of the structurally elevated smooth wear track and have the character of agglomerated and compacted debris in the rough parts of the track. The roughness of part of the wear track may be explained by similar arguments. Because it is only a glued agglomerate experiencing particularly high thermal gradients and low intrinsic strength, spallation due to hydrothermally induced pressure is a likely mechanism to keep it rough.

CONCLUSIONS

Wear in water lubricated SiC is viewed as a hydrothermal oxidation reaction in a dynamic situation. The conditions are such that there is an active type of corrosion, i.e. primary formation of silica as a layer is not expected from the static experiments and does not occur in the tribological tests. Instead the reaction produces a solution, from which silica precipitates, wherever the temperature or pressure falls.

The fluctuations are natural to the wear process and induce further erosion and fatigue phenomena, but help to create a tribolayer, which covers topographic features of the surface and consists of varying amounts of silicon carbide debris, ranging from micro- to nano-scaled grains, which are glued together by precipitated silica.

Further experimental studies combining static hydrothermal oxidation and tribological tests may allow building a quantitative wear theory capable of predicting wear rates.

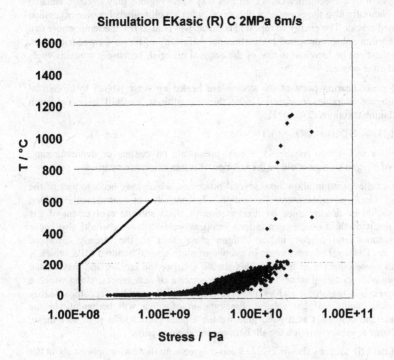

Fig. 1: Simulation of P-T conditions at contact points from a mechanical model (blue points) and a typical isochoric water heating model (red line)

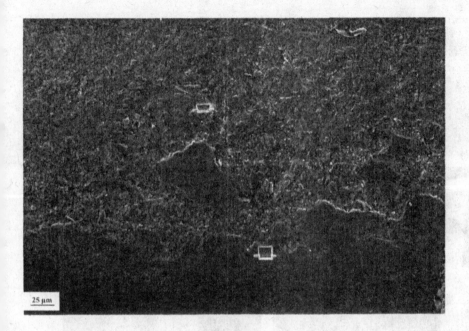

Fig. 2: REM graph of the worn SSiC sample with the smooth and rough part of the wear track, showing the location of the FIB sampling sites

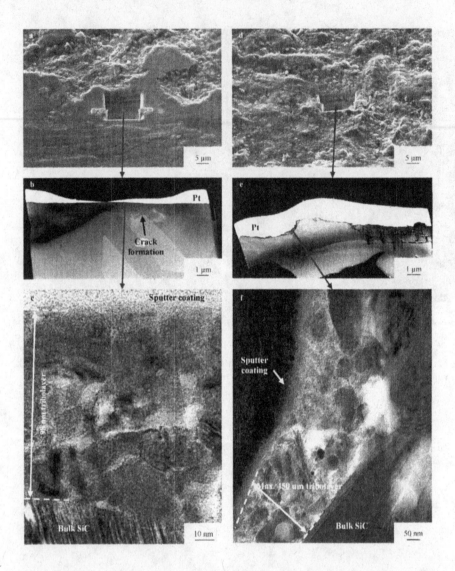

Fig. 3: Electron microphotographs of the superficial area of EKasic ® F wear tracks (a, d: SEM; b, e: STEM; c, f: TEM). The net-like structure on Fig. 3b and 3e originates from the TEM sample holder foil while white spots on Fig. 3e are due to Ga contamination.

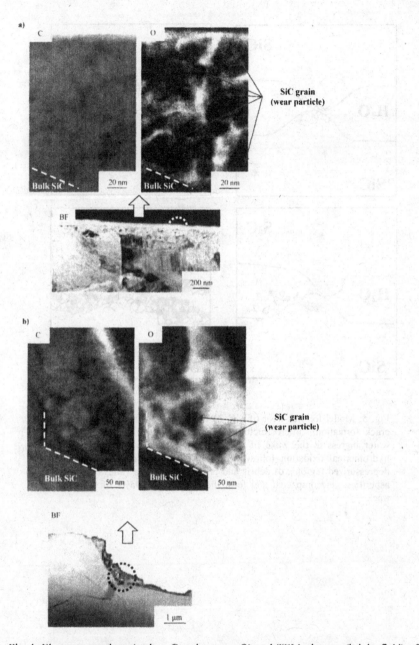

Fig. 4: Element mappings (carbon C and oxygen O) and TEM pictures (bright field) of (a) smoothened and (b) rough area on a EKasic ® F wear track. Between and kept together by an oxygen-rich matrix single rounded SiC grains (= transported wear particles) can be seen.

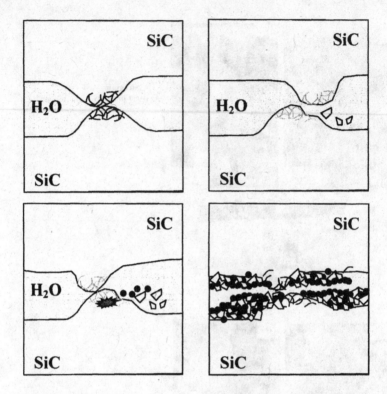

Fig. 5: Model for wet wear of SiC. a) first contact of asperities induces crack formation, microstructural grain diassemblage and spalling, b) water ingress in the wake of it, c) further collisions induce heat and hydrothermal oxidation followed by silica precipitation in cooling or depressurized regions, d) debris and precipitates fill up valleys between asperities, get compacted and form the smooth tribo-layer of the wear track

REFERENCES

1. Zum Gahr, K.-H., et al., *Micro- and macro-tribological properties of SiC ceramics in sliding contact.* Wear, 2001. **250**: p. 299-310.
2. Zhou, S. and H. Xiao, *Tribo-chemistry and wear map of silicon carbide ceramics.* Journal of the Chinese Ceramic Society, 2002. **30**(5): p. 641-644.
3. Cho, S.-J., C.-D. Um, and S.-S. Kim, *Wear and Wear Transition in Silicon Carbide Ceramics during Sliding.* Journal of the American Ceramic Society, 1996. **79**(5): p. 1247-1251.
4. Borrero-Lopez, O., et al., *Effect of Microstructure on Sliding-Wear Properties of Liquid-Phase-Sintered α-SiC.* Journal of the American Ceramic Society, 2005. **88**(8): p. 2159 - 2163.
5. Andersson, P., *Water-lubricated pin-on-disc tests with ceramics.* Wear, 1992. **154**: p. 37-47.
6. Andersson, P., et al., *Influence of topography on the running-in of water-lubricated silicon carbide journal bearings.* Wear, 1996. **201**: p. 1-9.
7. Chen, M., K. Kato, and K. Adachi, *Friction and wear of self-mated SiC and Si₃N₄ sliding in water.* Wear, 2001. **250**: p. 246-255.
8. Kailer, A., et al., *Tribologisches Verhalten von keramischen Gleitringdichtungen.* Tribologie + Schmierungstechnik, 2003. **50**: p. 10-13.
9. Quinn, T.F.J., *Oxidational wear modelling: I.* Wear, 1992. **153**(1): p. 179 - 200.
10. Quinn, T.F.J., *Oxidational wear modelling: Part II. The general theory of oxidational wear.* Wear, 1994. **175**(1): p. 199 - 208.
11. Xu, J. and K. Kato, *Formation of tribochemical layer of ceramics sliding in water and its role for low friction.* Wear, 2000. **245**: p. 61-75.
12. Alexander, G.B., W.M. Heston, and R.K. Iler, *The Solubility of Amorphous Silica in Water.* Journal of Physical Chemistry, 1954. **58**(1): p. 453 - 455.
13. Dove, P.M. and J.D. Rimstidt, *Silica-Water Interactions*, in *Silica*, P.J. Heaney, C.T. Prewitt, and G.V. Gibbs, Editors. 1994, Mineralogical Society of America: Washington. p. 259-308.
14. Perera, G., R.H. Doremus, and W. Lanford, *Dissolution Rates of Silicate Glasses in Water at pH 7.* Journal of the American Ceramic Society, 1991. **74**(6): p. 1269-1274.
15. Andersson, P., A.-P. Nikkilä, and P. Lintula, *Wear characteristics of water-lubricated SiC journal bearings in intermittent motion.* Wear, 1994. **179**: p. 57-62.
16. Morey, G.W., R.O. Fournier, and J.J. Rowe, *The solubility of quartz in water in the temperature interval from 25° to 300°C.* Geochimica et Cosmochimica Acta, 1962. **26**(1): p. 1029 - 1043.
17. Rimstidt, J.D. and H.L. Barnes, *The kinetics of silica-water reactions.* Geochimica et Cosmochimica Acta, 1980. **44**: p. 1683-1699.
18. Fournier, R.O. and J.J. Rowe, *The Solubility of Amorphous Silica in Water at high Temperatures and High Pressures.* American Mineralogist, 1977. **62**: p. 1052-1056.
19. Yoshimura, M., J.-I. Kase, and S. Somiya, *Oxidation of SiC powder by high-temperature, high-pressure H2O.* Journal of Materials Research, 1986. **1**(1): p. 100 - 103.
20. Yoshimura, M., et al., *Oxidation Mechanism of Nitride and Carbide Powders by High-Temperature High-Pressure Water*, in *Corrosion and Corrosive Degradation of Ceramics*, R.E. Tressler and M. McNallan, Editors. 1990, Am.Ceram.Soc. p. 337-354.
21. Gogotsi, Y.G. and M. Yoshimura, *Degradation of SiC (Tyranno) fibres in high-temperature, high-pressure water.* Journal of Materials Science Letters, 1994. **13**: p. 395-399.
22. Gogotsi, Y.G. and M. Yoshimura, *Formation of carbon films on carbides under hydrothermal conditions.* Nature, 1994. **367**: p. 628 - 630.

23. Gogotsi, Y.G., et al., *Structure of Carbon Produced by Hydrothermal Treatment of b-SiC Powder.* Journal of Materials Chemistry, 1996. **6**(4): p. 595-604.

24. Gogotsi, Y.G., et al., *Formation of sp³-bonded Carbon upon Hydrothermal Treatment of SiC.-.* Diamond and Related Materials, 1996. **5**(2): p. 151-162.

25. Kraft, T., K.G. Nickel, and Y.G. Gogotsi, *Hydrothermal Degradation of CVD SiC Fibers.* Journal of Materials Science, 1998. **33**: p. 4357-4364.

26. Kraft, T. and K.G. Nickel, *Carbon formed by hydrothermal treatment of a-SiC crystals.* Journal of Materials Chemistry, 2000. **10**(3): p. 671-680.

27. Barringer, E., et al., *Corrosion of CVD Silicon Carbide in 500°C Supercritical Water.* Journal of the American Ceramic Society, 2007. **90**(1): p. 315 - 318.

28. Kim, W.-J., et al., *Corrosion behaviors of sintered and chemically vapor deposited silicon carbide ceramics in water at 360°C.* Journal of Materials Science Letters, 2003. **22**: p. 581-584.

29. Kitaoka, S., et al., *Tribological Characteristics of SiC Ceramics in High-Temperature and High-Pressure Water.* Journal of the American Ceramic Society, 1994. **77**(7): p. 1851-1856.

30. Kitaoka, S., et al., *Tribochemical wear theory of non-oxide ceramics in high-temperature and high-pressure water.* Wear, 1997. **205**: p. 40-46.

31. Tworzydlo, W.W., et al., *Computational micro- and macroscopic models of contact and friction: formulation, approach and applications.* Wear, 1998. **220**: p. 113-140.

32. Bhushan, B., *Contact mechanics of rough surfaces in tribology: multiple asperity contact.* Tribology Letters, 1998. **4**(1): p. 1 - 35.

33. Kuhlmann-Wilsdorf, D., *Demystifying Flash Temperatures. I. Analytical Expressions based on a simple model.* Materials Science and Engineering, 1987. **91**(1): p. 107 - 118.

34. Bassett, W.A., et al., *A new diamond anvil cell for hydrothermal studies to 2.5 GPa and from -190 to 1200°C.* Review of Scientific Instruments, 1993. **64**(8): p. 2340 - 2345.

35. Presser, V., et al., *Tribological and hydrothermal behaviour of silicon carbide under water lubrication.* Wear, submitted.

36. Haselton, H.T., et al., *Techniques for determining pressure in the hydrothermal diamond-anvil cell: Behaviour and identification of ice polymorphs (I, II, V, VI).* American Mineralogist, 1995. **80**(11 - 12): p. 1302 - 1306.

37. Shen, A.H., W.A. Bassett, and I.-M. Chou, *Hydrothermal Studies in a Diamond Anvil Cell: Pressure Determination Using the Equation of State of H₂O*, in *High-Pressure Research: Application to Earth and Planetary Sciences*, Y. Syono and M.H. Manghnani, Editors. 1992, American Geophysical Union: Washington D.C. p. 61 - 68.

38. Wirth, R., *Focused ion beam (FIB): a novel technology for advanced application of micro- and nanoanalysis in geosiences and applied mineralogy.* European Journal of Mineralogy, 2004. **16**(6): p. 863-867.

Reliability, NDE, and Fractography

PROBABILISTIC DESIGN OPTIMIZATION AND RELIABILITY ASSESSMENT OF HIGH TEMPERATURE THERMOELECTRIC DEVICES

O. M. Jadaan
College of Engineering, Mathematics, and Science
University of Wisconsin-Platteville
Platteville, WI 53818

A. A. Wereszczak
Ceramic Science and Technology
Oak Ridge National Laboratory
Oak Ridge, TN 37831

ABSTRACT

Thermoelectric (TE) devices, subcomponents of which are made of brittle materials, generate an electric potential when they are subjected to thermal gradients through their thickness. These devices are of significant interest for high temperature environments in transportation and industrial applications where waste heat can be used to generate electricity (also referred to as "waste heat recovery" or "energy harvesting"). TE devices become more efficient as larger thermal gradients are applied across them. This is accomplished by larger temperature differences across the TE's hot and cold junctions or the use of low thermal conductivity TE materials or both. However, a TE brittle material with a combination of poor strength, low thermal conductivity, and large coefficient of thermal expansion can translate into high probability of mechanical failure (low reliability) in the presence of a thermal gradient, thereby preventing its use as intended. The objective of this work is to *demonstrate* the use of an established probabilistic design methodology developed for brittle structural components and corresponding design sensitivity analyses to optimize the reliability of an arbitrary TE device. This method can be used to guide TE material and design selection for optimum reliability. The mechanical reliability of a prototypical TE device is optimized from a structural ceramic perspective, using finite element analysis and the NASA CARES/Life integrated design code. Suggested geometric redesigns and material selection are identified to enhance the reliability of the TE device.

INTRODUCTION

Thermoelectric (TE) modules are solid-state heat pumps that operate on the Peltier effect [1]. A TE module consists of an array of p- and n- type semiconductor elements heavily doped with electrical carriers. The array of elements is soldered so that it is electrically connected in series and thermally connected in parallel [1]. This array is then often affixed to two ceramic substrates (insulators), one on each side of the elements as seen in Figure 1.

Potential next generation thermoelectric (TE) devices comprised of p- and n-type oxide ceramics enjoy strong interest for implementation in high temperature and oxidizing environments because their waste heat could be used to generate electricity. However, the intended TE function of these devices will only be enabled if the device is designed to overcome the thermomechanical limitations (i.e., brittleness) inherent to these oxides. A TE oxide with a combination of poor strength, low thermal conductivity, and large coefficient of thermal expansion can readily fail in the presence of a thermal gradient thereby preventing the exploitation of the desired thermoelectrical function.

This problem can be overcome with the combined use of established probabilistic design methods developed for brittle structural components, good thermoelastic and thermomechanical databases of the candidate oxide material comprising the TE device, and iteratively applied design sensitivity analysis. Therefore, the objective of this work is to *demonstrate* the use of a probabilistic design methodology, whereby established probabilistic design methods developed for brittle structural components and iteratively applied component design sensitivity analyses are utilized, to optimize the reliability of an arbitrary TE device.

There are several outcomes from this work that can benefit the TE device developers and end-users of these potentially high temperature TE devices: mechanical reliability of prototypical TE devices are evaluated from a structural ceramic perspective and suggested redesigns are identified. These redesigns include material selection as well as geometric dimensions that would minimize the stresses and failure probabilities in the TE and ceramic substrate materials.

PROBABILISTIC DESIGN OPTIMIZATION APPROACH

Brittle materials display stochastic strength behavior due to their low fracture toughness and the random nature of inherent microscopic flaw sizes, orientation and distribution. This dispersion in strength requires a probabilistic life prediction and design methodology. The NASA-developed CARES/Life code [2] predicts the probability of failure for ceramic structures subjected to transient multiaxial thermomechanical loading. This code models the material strength as the only probabilistic quantity (random input variable). Such analysis yields a conditional probability of failure based on the condition that temperature and stress fields within the component are deterministic and thus not affected by random variations in geometric tolerances, load histories, and scatter in material parameters.

However, in many applications the variability of parameters other than strength can be significant and must be taken into account. The CARES/Life code has been coupled to the ANSYS Probabilistic Design System (PDS) to consider the total probability of failure using the entire space of random input variables [3-5]. The ANSYS-PDS is a probabilistic design software integrated within the ANSYS Finite Element Analysis (FEA) program. When coupled with the CARES/Life program, PDS computes the total probability of failure by accounting for uncertainty in the component's dimensions, loading, and material properties by assigning statistical distributions to these random input parameters and performing Monte Carlo like simulation methods. This CARES-PDS capability would enable more realistic assessment of brittle material structural integrity. The theoretical background to this total probabilistic design approach is outlined in references [3] and [4] and will not be repeated here.

This paper demonstrates the application of the total probabilistic design environment to the design and development of a TE module to be used for energy harvesting. Material and geometric variabilities are considered, including the Weibull parameters themselves. The probabilistic design optimization is performed in two general steps:

1) Use the ANSYS PDS capability to select a material system and geometric dimensions to minimize stresses in the ceramic substrate (insulation plates) and TE brittle materials. A material system is composed of the downselected materials for the TE, ceramic substrate, and electric contact materials.

2) For the downselected material system and TE device geometry, investigate the influence of material and geometric tolerances (variability) on the total probability of failure and determine which parameters have the most impact on the reliability of the device. This step involves using both the ANSYS PDS and the CARES/Life codes. The CARES code is used in this step since the probability of failure for the TE device is to be computed.

FEA SIMULATION FOR A THERMOELECTRIC DEVICE

FEA stress analysis for a TE module is performed first. Subsequent PDS analysis is conducted whereby 11 geometric and material parameters are varied to minimize two response variables, the maximum principal stress in the TE legs (S1_TE) and the maximum principal stress in the ceramic substrates or insulator plates (S1_insulator). This initial PDS study yields sensitivity plots showing which random input variables most affect the stresses in the TE and insulator materials. In response, a material system and geometric dimensions are specified for the optimized device.

Figure 1 shows the geometry of the TE module with colors corresponding to the initially selected material system. This design and material system shown in this figure reflects a typical existing design to be optimized. In this schematic, purple corresponds to alumina insulator plates, turquoise for steel contacts, while red is for the TE legs (made from Skutterudite in this analysis). Figure 2 shows various views of the TE module with the insulator plates removed in order to clearly display the geometric arrangement of the steel contacts and TE legs.

Table I lists the dimensions of the various components making up the initial design for the TE device. Table II contains the thermomechanical properties for the three materials involved in the module. To keep the demonstration of the proposed design methodology simple, these properties are assumed to be temperature independent. However, taking into account temperature dependent material properties can be easily incorporated into the FEA simulation and the probabilistic design approach.

Table I - Dimensions of TE module

Dimension	Magnitude (mm)
Alumina insulator plate width	44
Top Alumina insulator plate length	135
Bottom Alumina insulator plate length	128
Alumina insulator plate thickness	2
TE leg width	5
TE leg height	7
Contact thickness	1.5
Spacing between contacts	2

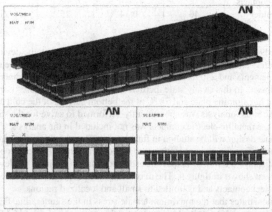

Figure 1 - Various views of the initial design for the TE module with colors corresponding to the following material system: Purple is the alumina insulator plates, turquoise is the steel contacts, while red is the Skutterudite TE legs.

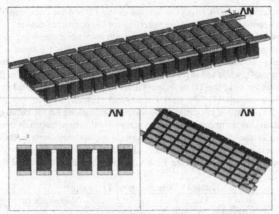

Figure 2 - Various views of the TE module with the insulator plates removed in order to clearly display the geometric arrangement of the TE legs and the steel contacts.

Table II - Thermomechanical material properties

Materials	E (GPa)	Poisson's ratio	K (W/m.k)	CTE Ppm/°C
Skutterudite TE	24.5	0.267	5	18
Alumina Plates	375	0.22	35	8
410S stainless steel contacts	200	0.33	24.9	12
Solder	40	0.34	25	21

Thermal solid90 and structural solid95 elements were used to mesh the device and conduct the thermomechanical stress analysis. Figure 3 shows the mesh distribution used in the simulation containing 142899 elements and 240989 nodes. Figure 4 shows the resulting temperature distribution throughout the TE device. In this steady state thermal analysis, the temperatures were set to 533 °C at the top surface of the hot alumina plate and 78 °C at the bottom surface of the cold alumina plate.

Thermoelastic stress analysis was subsequently performed to solve for the stresses within the device. Plasticity in the metallic electric contacts was not included in the analysis. The effect of yielding on the module design will be studied in future work. This should not be construed as a drawback to this work, since the main objective is to demonstrate the proposed probabilistic design approach. Figure 5 displays the first principal stress distribution in the entire TE device due to the 455 °C temperature gradient shown in figure 4. The maximum stress that develops in the module is located within the stainless steel contacts and is limited to small and localized regions.

Figure 6 demonstrates that the maximum tensile stress in the skutterudite TE material reaches 382 MPa in localized areas, while in the Alumina plates the stress goes up to 815 MPa as seen in figure 7. Given that the strength of Skutterudite is estimated to be about 50 MPa, while that for Alumina is roughly 500 MPa, it can be seen that this design will not survive the intended service temperature shown in figure 4.

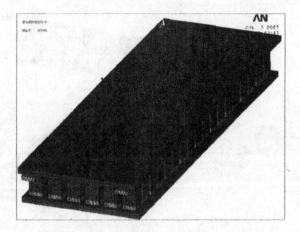

Figure 3 - Mesh distribution showing solid 95 elements. The mesh contains 142899 elements and 240989 nodes.

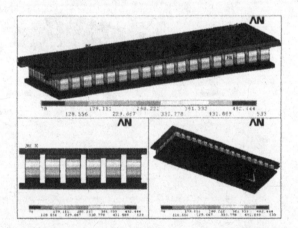

Figure 4 - Temperature distribution in the TE module in units of degrees Celsius.

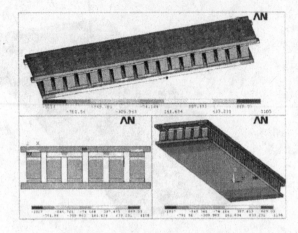

Figure 5 - First principal stress distribution in the TE device due to the thermal load shown in figure 4.

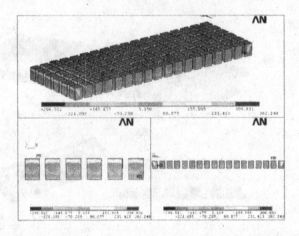

Figure 6 - First principal stress distribution in the TE legs due to the thermal load shown in figure 4.

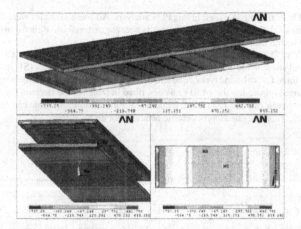

Figure 7 - First principal stress distribution in the Alumina insulation plates due to the thermal load shown in figure 4.

DESIGN ANALYSIS TO SELECT MATERIAL SYSTEM AND GEOMETRIC DIMENSIONS FOR THERMOELECTRIC DEVICE

Thermal stresses develop within the TE module due to thermal mismatches between the various materials making up the device. Hence, the goal of this portion of the PDS analysis is to select a material system and geometric dimensions that would minimize the maximum tensile stresses in the materials making up the TE legs and ceramic substrate.

The PDS capability permits performing Monte Carlo simulations with multiple input random variables. Geometric, material, and load parameters can be varied by assigning statistical distribution functions to them. Response variables are defined so that the device can be optimized by minimizing or maximizing them. The PDS analysis can then be used to perform sensitivity analysis to determine which random input variables should be altered and how, in order to optimize the selected response variables.

In this analysis, two response random variables were defined. These are:

1) Maximum principal stress in the TE legs (S1_TE)
2) Maximum principal stress in the Alumina insulation plates (S1_insulator)

The goal is to determine which geometric and/or material parameters have the most influence on S1_TE and S1_insulator.

Fourteen geometric and material parameters were assumed to be random input variables (RIV) and are listed in table III. For example, TE leg and insulator plate thicknesses were assumed to vary within ranges limited by geometric design constraints. For the TE design of figure 1, The TE leg height must be between 2 and 10 mm, while the insulator plate thickness is constrained to within 1.5 mm and 4 mm. The material properties for the TE, substrate, and electric contact materials are also varied within limits representing the range of available materials that can be used for the components of the TE device. For example, the range of materials that can be used to make the TE legs have elastic moduli ranging between 25 GPa and 125 GPa, while their thermal conductivities are within 3 and 25

W/m.K. Table III lists all the RIVs used in this PDS analysis. All these variables were defined to vary using uniform statistical distribution functions. This is because material and dimensional selection is equally weighted within the ranges specified in table III.

The ANSYS PDS analysis was performed using 100 samples. Figures 8a nd 8b show the maximum principal stress in the TE legs (S1_TE) and alumina insulation plates (S1_insulator) as function of 100 Monte Carlo-Latin Hyper cube simulations, respectively. As can be seen from these two figures and based on varying the 14 RIVs as described in table III, the stress in the TE legs can range anywhere between 180 and 1533 MPa, while that for the ceramic substrate can range between 203 and 2356 MPa. This analysis assumes elastic behavior for all materials.

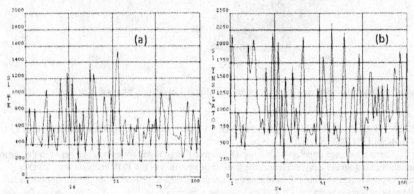

Figure 8 - Maximum principal stress in: (a) TE legs (S1_TE) and (b) alumina insulation plates (S1_insulator) as function of the 100 Monte Carlo-Latin Hyper cube simulations.

Figure 9 is the sensitivity plot highlighting the relative importance of the random input variables on S1_TE. In this figure, a positive correlation for a given RIV means that as this variable increases so does S1_TE, and vice versa. The following can be concluded for figure 9:

1) S1_TE decreases as the coefficient of thermal expansion (CTE) for the TE material increases and CTE for the contact material decreases. In general, the TE CTE (ranges between 6 and 24 ppm/C) is less than the contact CTE (ranges between 12 and 30 pp/C). *Therefore the tendency is for the stress in the TE legs to decrease by decreasing the thermal mismatch between the TE legs and the contact pads.*
2) S1_TE decreases as the TE thermal conductivity increases.
3) S1_TE decreases as the elastic moduli for both the TE and contact materials decrease.
4) The geometric variables considered in this analysis (see table III) have negligible effect on S1_TE compared to the material variables described above

Table III - Specifications for the random input variables used in this study. Material designations are: 1 = contact material; 2 = insulation or substrate material; 3 = TE material.

Random input variable	Statistical distribution	Minimum value	Maximum value	comments
Material Random Variables				
E_3 (GPa)	uniform	25	125	Representing TE materials with different Moduli
n_3	uniform	0.2	0.3	Representing TE materials with different Poisson's ratios
K_3 (W/m.K)	uniform	3	25	Representing TE materials with different thermal conductivities
CTE_3 (ppm/C)	uniform	6	24	Representing TE materials with different CTEs
E_2 (GPa)	uniform	350	400	Representing insulation materials with different Moduli
n_2	uniform	0.2	0.25	Representing insulation materials with different Poisson's ratios
K_2 (W/m.K)	uniform	20	50	Representing insulation materials with different thermal conductivities
CTE_2 (ppm/C)	uniform	7	9	Representing insulation materials with different CTEs
E_1 (GPa)	uniform	10	200	Representing contact materials with different Moduli
K_1 (W/m.K)	uniform	20	60	Representing contact materials with different thermal conductivities
CTE_1 (ppm/C)	uniform	12	30	Representing contact materials with different CTEs
Geometric Random Variables				
LH (mm)	uniform	2	10	Range of heights for TE leg
LWX (mm)	uniform	3	6	Range of widths for TE legs
IH (mm)	uniform	1.5	4	Range of thicknesses for insulation plates

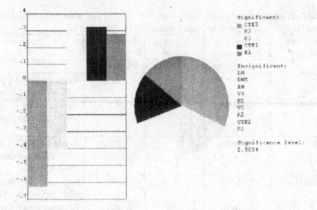

Figure 9 - Sensitivity plot showing random variables with highest impact on the maximum principal stress in the TE legs (S1_TE). Material designations are: 1 = contact material; 2 = insulation or substrate material; 3 = TE material.

Therefore, to reduce the maximum stress in the TE material, three things can be done:

1) Select TE and contact materials with minimal CTE mismatch
2) Select TE material with high thermal conductivity
3) Select soft TE and contact materials (low elastic moduli)

Figure 10 is the sensitivity plot highlighting the relative importance of the random input variables on S1_insulator. The following can be concluded:

1) S1_insulator increases as the elastic modulus for the contact material increases. Hence, select soft contact material.
2) S1_insulator increases as CTE for the contact material increases since this causes the CTE mismatch between the contact material and substrate to increase. Therefore select a contact material with low CTE but also with minimal CTE mismatch relative to that for the TE material
3) S1_insulator increases as the TE leg height decreases. Hence, maximize height of TE legs to decrease the stress in the substrate.

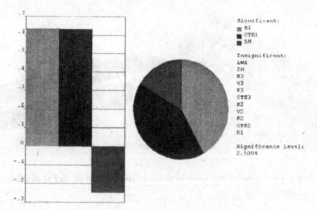

Figure 10 - Sensitivity plot showing random variables with highest impact on the maximum principal stress in the Alumina insulation plates (S1_insulator).

CONDITIONAL PROBABILITY OF FAILURE FOR THE REDESIGNED THERMOELECTRIC DEVICE:

The traditional approach for determining the reliability of ceramic materials assumes the FEA model (load, geometry, and material properties) to be deterministic and the strength to be the only random input variable. This makes the probability of failure (Pf) a conditional probability, since it is based on the condition that load, geometry and material parameters have no uncertainty associated with them. In reality, however, all these parameters are subject to scatter. Taking all variabilities into account can significantly change the predicted probability of failure [2-5]. In this section, the conditional probability of failure for a redesigned TE device is computed using Weibull theory and the CARES/Life code [2].

The TE module was redesigned using the PDS analytical outcomes described in the previous section. In this redesign the substrate and the TE materials were kept the same. Furthermore, the geometrical shape was not to be altered but the dimensions could be changed. The following changes were built-into into the model:

1) *Replace steel contacts with solder material* (see table II for properties). Selecting the solder material satisfies two of the recommendations that: a) softer contact material be used, and b) reduce CTE mismatch between the contact and TE materials.

2) *Increase TE leg heights* in order to reduce stress in the substrate material. The leg heights were increased from 7 mm to 10 mm which is the maximum dimension permitted per table III.

3) The Skutterudite TE and alumina ceramic substrate materials were not changed.

Figure 11 displays the principal stresses in the Skutterudite TE legs and alumina substrate. Comparing this figure to figures 6 and 7, it can be seen that the stress in the TE material is reduced from 382 MPa to 283 MPa (26% reduction in S1_TE), while the stress in the alumina substrate decreased from 815 MPa to 668 MPa (18% reduction in S1_substrate).

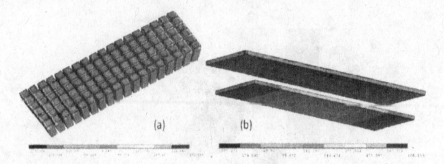

Figure 11 – First principal stress distribution in a) Skutterudite TE legs, and b) alumina substrate of the redesigned TE module.

The Weibull parameters for the Alumina and Skutterudite materials, required to compute the Pf, are listed in table IV. The Weibull multiaxial theory and details about how CARES/life computes the combined failure probability in both the substrate and TE brittle materials are well outlined in references [2-5] and thus will not be repeated in this paper because of space limitation.

Table IV – Weibull parameters for the alumina and Skutterudite materials

Material	Weibull modulus - m	Volume scale parameter – s_{0V} (MPa.mm3/m)
Alumina	15	500
Skutterudite	10	50

Because of the low strengths for both brittle materials compared to the applied thermal stresses, the predicted conditional probability of failure is predicted to be 100% for the redesigned module under the prescribed temperature gradient of 455 °C. It is apparent that the applied thermal load is rather excessive for this TE device and thus should be moderated for the device not to fail.

The analysis was redone assuming a thermal gradient equal to 20% of the 455 °C stated above or 91 °C. For this thermal load the max stresses in the TE legs and alumina substrate decreased to 39MPa and 189 MPa, respectively. These stresses resulted in conditional Pf for the TE legs and substrate materials to be 0.186 and 1.03e-7, respectively. The conditional Pf for the entire device (combining both TE and substrate materials) was 0.186, essentially equal to that of the TE material.

TOTAL PROBABILITY OF FAILURE AND SENSITIVITY ANALYSIS FOR THE REDESIGNED THERMOELECTRIC DEVICE:
Total probability of failure analysis takes into account the influence of multiple random input variables on the device's Pf. As was stated earlier, in many applications the variability of parameters other than strength can be significant and must be taken into account. The CARES/Life code has been coupled to the ANSYS Probabilistic Design System (PDS) to consider the total probability of failure using the entire space of random input variables [3-5]. When coupled with the CARES/Life program, PDS computes the total probability of failure by accounting for uncertainty in the component's

dimensions, loading, and material properties by assigning statistical distributions to these random input parameters and performing Monte Carlo like simulation methods.

The total Pf analysis to follow is based on the reduced thermal load gradient of 91 °C mentioned in the previous section which induces a conditional Pf of 0.186. Table V summarizes the 16 random variables, the statistical distribution functions assigned to them, and the corresponding distribution parameters reflecting their scatter. Some of the random parameters from the previous PDS analysis were removed since they did not strongly influence the maximum stresses in the TE and substrate materials. The four Weibull parameters (two Weibull moduli and two scale parameters) for the TE and substrate materials were assumed to be RIV with Gaussian distributions (see table V). The statistical distribution types used to describe these RIV are not based on data, but still realistically describe their uncertainty.

Table V - Specifications for the random input variables used in this study

Random input variable	Statistical distribution	Parameters
E_3 (GPa) – Skutterudite elastic modulus	Gaussian	Mean = 24.5 Standard dev = 0.1* mean
K_3 (W/m.K) – Skutterudite thermal conductivity	Gaussian	Mean = 5. Standard dev = 0.1* mean
CTE_3 (ppm/C) – Skutterudite CTE	Gaussian	Mean = 18 Standard dev = 0.1* mean
E_2 (GPa) – alumina elastic modulus	Gaussian	Mean = 375 Standard dev = 0.05* mean
K_2 (W/m.K) – alumina thermal conductivity	Gaussian	Mean = 35 Standard dev = 0.05* mean
CTE_2 (ppm/C) – alumina CTE	Gaussian	Mean = 8 Standard dev = 0.05* mean
E_1 (GPa) – solder elastic modulus	Gaussian	Mean = 40 Standard dev = 0.05* mean
K_1 (W/m.K) – solder thermal conductivity	Gaussian	Mean = 25 Standard dev = 0.05* mean
CTE_1 (ppm/C) – solder CTE	Gaussian	Mean = 21 Standard dev = 0.05* mean
m_2 – alumina Weibull modulus	Gaussian	Mean = 15 Standard dev = 4
s_{0V2} – alumina scale parameter	Gaussian	Mean = 500 Standard dev = 50
m_3 – Skutterudite Weibull modulus	Gaussian	Mean = 10 Standard dev = 3
s_{0V2} – Skutterudite scale parameter	Gaussian	Mean = 50 Standard dev = 10
LH (mm) – TE leg height	Gaussian	Mean = 10 Standard dev = 0.1* mean
LWX (mm) – TE leg width	Gaussian	Mean = 5 Standard dev = 0.1* mean
IH (mm) – substrate thickness	Gaussian	Mean = 2 Standard dev = 0.1* mean

Table VI lists the results for the three random response parameters which are the total probability of failure, the maximum principal stress in the Skutterudite TE legs, and the maximum principal stress in the alumina substrate. As can be seen from this table, the mean total probability of failure for the TE device is computed to be 0.391, the mean maximum principal stress in the TE legs equals 32 MPa, while the mean maximum principal stress in the substrate is 160 MPa. However, due to the geometric and material uncertainties listed in table V, the TE's total probability of failure could be anywhere between 0.00018 and 1.0. This means the maximum total probability of failure could be significantly higher than the 0.186 conditional probability of failure computed assuming all parameters other than strength to be deterministic. This result highlights the importance of taking into account all uncertainties thought to affect the reliability of the system.

Table VI- Statistics for the total probability of failure and maximum principal stresses in the TE and substrate materials as generated by ANSYS PDS and CARES/Life.

Name	Mean	Standard Deviation	Minimum	Maximum
Total Probability of failure	0.391	0.377	0.000180	1.000
Maximum stress in the TE Material (MPa)	32.0	8.0	18.7	57.5
Maximum stress in the substrate (MPa)	159.6	13.9	117.0	196.0

In the PDS analysis described above, 94 simulation loops (samples) were used to perform the Monte Carlo simulation. The reason 94 instead of 100 samples were used in this analysis is because 6 of the 100 FEA simulations failed to solve and thus were removed from the final results. Figure 12 shows the sample history plots for the total probability of failure. Figure 12a displays the TE device Pf for the 94 different simulations, clearly showing how dependent the Pf is on material variability and geometric tolerance. In figure 12b, the narrowing 95% confidence bounds for the mean Pf indicate that enough samples were used for this analysis. Of course more samples would yield tighter confidence bounds but at the expense of solution time. Such figures can be used to assess the quality of results for the output response variables based on the number of simulation loops.

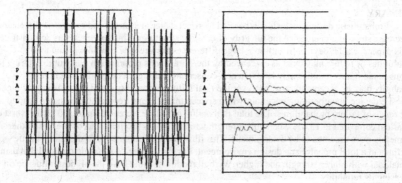

Figure 12 - Sample history plots for the total probability of failure. (a) – Probability of failure history (b) – Mean value with 95% confidence interval history

An important outcome to designing using a total probabilistic design environment is determining which random variables have the most impact on the structure's probability of failure for efficient focus of resources (sensitivity analysis). Figure 13 is a sensitivity plot highlighting which random variables have the greatest impact on the device's Pf. It can be seen from this figure that only three out of the 16 prescribed random input variables significantly affect the P_f. These are the Weibull modulus, the scale parameter, and the elastic modulus of the TE material in that order. As these values increase, the Pf for the TE device decreases. These results make sense because almost the entire Pf in the device is due to failure in the TE material. Therefore, using stronger TE material with high Weibull modulus would result in higher reliability for the TE module.

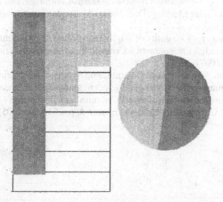

Figure 13 - Sensitivity plot for the probability of failure of the TE device using 94 Monte Carlo simulations. The significant parameters influencing the Pf in order of importance are: Weibull modulus of the TE material, scale parameter of the TE material, and elastic modulus of the TE material.

SUMMARY

Probabilistic design analysis can be used to select material systems and geometric dimensions to decrease stresses and enhance the integrity of ceramic structures. This approach was demonstrated in this paper for a TE device. In addition, large differences between the conditional and total probabilities of failure may exist as evidenced by the TE analysis presented in this work. Ignoring the various random uncertainties may lead to a non-conservative design. The conditional failure probability may be acceptably low, however the total failure probability could be significantly higher. In the TE device, incorporating uncertainty into sixteen material and geometric parameters lead to a wide range of failure probabilities. The total probability of failure increased up to 100% (for worst case scenarios) compared to 18.6% when only strength variability was taken into account and all other parameters were assumed to be deterministic. This illustrates the importance of designing brittle structures in a total probabilistic design environment. For the TE device it was determined that using a TE material with higher strength and higher Weibull modulus would have the highest influence on increasing its reliability.

ACKNOWLEDGEMENTS

Research sponsored by the U.S. Department of Energy, Assistant Secretary for Energy Efficiency and Renewable Energy, Office of Vehicle Technologies, as part of the Propulsion Materials Program, under contract DE-AC05-00OR22725 with UT-Battelle, LLC. The authors wish to thank Connecticut Reserve Technologies for providing them with a free license of the CARES/Life code.

REFERENCES

[1] http://www.tetech.com/.

[2] N. Nemeth, O. Jadaan, and J.P. Gyekenyesi, "Lifetime Reliability Prediction of Ceramic Structures Under Transient Thermomechanical Loads," NASA TP-2005-212505 (2005).

[3] S. Reh, T. Palfi, and N. Nemeth, "Probabilistic Analysis Techniques Applied to Lifetime Reliability Estimation of Ceramics," Paper No. APS-II-49 Glass. JANNAF 39th Combustion Subcommittee, 27th Airbreathing Propulsion, 21st Propulsion Systems Hazards Committee, and 3rd Modeling and Simulation Subcommittee Joint Meeting (Dec. 2003). Available from Chemical Propulsion Information Agency (CPIA).

[4] O. Jadaan, and J. Trethewey, "Reliability of High Temperature Lightweight Valve Train Components in a Total Probabilistic Design Environment," Ceramic Engineering and Science Proceedings, Volume 27, Issue 2, p. 533-542 (2007).

[5] R. Carter, and O. Jadaan, "The Effects of Incorporating System Level Variability into the Reliability Analysis for Ceramic Components," Ceramic Engineering and Science Proceedings, Proceedings of the 29th International Conference on Advanced Ceramics and Composites - Developments in Advanced Ceramics and Composites, Volume 26, Issue 8, p. 253-260 (2005).

DEVELOPMENT OF A NEW COMPUTATIONAL METHOD FOR SOLVING INHOMOGENEOUS AND ULTRA LARGE SCALE MODEL

H. Serizawa, A. Kawahara, S. Itoh and H. Murakawa
Joining and Welding Research Institute, Osaka University
11-1 Mihogaoka, Ibaraki, Osaka 567-0047, Japan

ABSTRACT

The finite element method is a powerful tool to predict not only the mechanical behavior of structures but also the residual stresses and distortions caused in the fabrication processes. In order to solve ultra large finite element problems, the authors have proposed the Fractal Multi-Grid Method. In this method, the domain to be analyzed is subdivided into a multi-grid which has fractal or hierarchical structure and the solution is obtained by solving small cells at each hierarchy successively while allowing discontinuity between neighboring cells. The continuity is retained through iterations. In this research, a new type of Hierarchical Multi-Grid Method which satisfies the continuity is developed and its superiority over the old type Fractal Multi-Grid Method developed by the authors is demonstrated.

INTRODUCTION

As a result of R & D efforts on the computer technology in both hardware and software, it becomes possible to solve large scale finite element models with more than one hundred million degrees of freedom. As for the software improvement, various methods for solving large classes of matrix equations have been developed to decrease the computational time. These are, for example, ICCG (Incomplete Cholesky Conjugate Gradient) method [1], multifrontal method [2], multi-grid method [3] and AMG (algebraic multi-grid) method [4]. Also, DDM (Domain Decomposition Method) [5] and BDD (Balancing Domain Decomposition) method [6] have been developed for the parallel and grid computing. However, to compute the stress of complex real ceramic and ceramic composite materials with reasonable accuracy, the size of the finite element model becomes much larger. In such a case, a great amount of computational time and memory are necessary although the capacity of computer has been improved drastically. Thus, it is necessary to speed up the computation and increase its efficiency greatly.

To realize the stress analysis of geometrically complex model such as the advanced ceramic and composite materials, a program of the Fractal Multi Grid (FMG) method has been developed by the authors [7,8]. The idea of FMG method can be illustrated using a two dimensional simple elastic problem as shown in Fig. 1. A square sheet is stretched at its four vertexes. When the model is subdivided into 8 × 8 elements, the deformation and the stress are computed by solving basic cells consist of 2 × 2 elements under the prescribed displacements at four vertexes. Such basic operation is repeated hierarchically from the top level to the lowest level. In this process, the continuity between the neighboring cells is ignored and it can be recovered at the lowest level by interpolating the displacements at two nodes sharing the same cell boundary. In this way, the continuity of the traction is not guaranteed. It is retained through the iteration. Also, the authors have demonstrated the potential effectiveness of the Fractal Multi-Grid Method. At the same time, intrinsic problem associated with the convergence was found. The convergence can be maintained with controlling the iterative procedure by reducing the feedback ratio. The appropriate value of the feedback ratio depends on the problem and the order of hierarchy. The reason for this problem is that this type of FMG method does not have a solid theoretical background such as the Minimum Potential Energy Theorem. Therefore, the authors attempted to establish its theoretical bases on the Principle of Minimum Potential Energy.

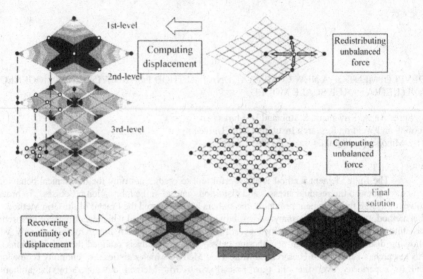

Fig. 1   Procedure of computation in FMG method.

THEORY OF HIERARCHICAL MULTI-GRID METHOD

As in the ordinary finite element method (FEM), the proposed Hierarchical Multi-Grid Method can be constructed on the well known Principle of Minimum Potential Energy, i.e.

$$\Pi(u) = \frac{1}{2} \int [\sigma][\varepsilon] dv - \int [g][u] dv - \int [\tau][u] ds_\sigma \tag{1}$$

where, u : displacement, σ : stress, ε : strain, g : body force, τ : traction applied as the external load. In the present HMG, the displacement u is interpolated by the function with hierarchical structure, i.e.

$$u = \sum_{h=1}^{H} u_h = \sum_{h=1}^{H} [A_h][U_h] \tag{2}$$

where, $[A_h]$ is the interpolation function and $\{U_h\}$ is the nodal displacement of the nodes belonging to the h-th hierarchy. Using Eq. (2), the strain-displacement relation and the stress-strain relation can be described in the following form.

$$\{\varepsilon\} = \sum_{h=1}^{H} [B_h][U_h] \tag{3}$$

$$\{\sigma\} = [D]\{\varepsilon\} = [D] \sum_{h=1}^{H} [B_h][U_h] \tag{4}$$

Substituting Eqs. (3) and (4) into Eq. (1), the functional $\Pi(u)$ can be rewritten in terms of the nodal displacement parameter $\{U_h\}$ with hierarchical structure.

$$\Pi(U_1,\cdots,U_h,\cdots,U_H) = \frac{1}{2}\int \left(\sum_{h=1}^{H}[U_h][B_h]\right)' [D]\left(\sum_{h=1}^{H}[B_h]\{U_h\}\right)dv - \int [g]\left(\sum_{h=1}^{H}[A_h]\{U_h\}\right)dv$$
$$- \int [\tau]\left(\sum_{h=1}^{H}[B_h]\{U_h\}\right)ds_\sigma \tag{5}$$

Based on the stationality condition of the above functional, an iterative solution procedure can be constructed. Let us assume, that $\{U_1,\cdots,U_h,\cdots,U_H\}$ is the current approximation and $\{\Delta U_h\}$ is the correction for $\{U_h\}$. The correction vector $\{\Delta U_h\}$ can be obtained through the stationality condition of the following functional.

$$\Pi(U_1,\cdots,U_h+\Delta U_h,\cdots,U_H) = \frac{1}{2}\int [\Delta U_h][B_h]'[D][B_h]\{\Delta U_h\}dv + \int [\Delta U_h][B_h]'[D]\left(\sum_{h=1}^{H}[B_h]\{U_h\}\right)dv$$
$$- \int [\Delta U_h][A_h]'\{g\}dv - \int [\Delta U_h][B_h]'\{\tau\}ds_\sigma \tag{6}$$

Since the displacement parameter $\{U_h\}$ consists of a set of nodes (from 1 to N_h), the correction is made node by node, i.e.

$$\{U_h+\Delta U_h\} = [U_h^1, U_h^2,\cdots,U_h^n+\Delta U_h^n,\cdots,U_h^{N_h}]' \tag{7}$$

Thus, the stationality condition can be written as,

$$\delta\Pi(\delta\Delta U_h^n) = \int [\delta\Delta U_h^n][B_h^n]'[D][B_h^n]\{\Delta U_h^n\}dv + \int [\delta\Delta U_h^n][B_h^n]'[D]\left(\sum_{h=1}^{H}[B_h]\{U_h\}\right)dv$$
$$- \int [\delta\Delta U_h^n][A_h^n]'\{g\}dv - \int [\delta\Delta U_h^n][A_h^n]'\{\tau\}ds_\sigma = 0 \tag{8}$$

Since Eq. (8) must hold for arbitrary value of $\{\delta\Delta U_h^n\}$, the following equation is derived.

$$\{\Delta U_h^n\} = [K_h^n]^{-1}\{f_h^n\} \tag{9}$$

where, $\{f_h^n\}$ is the residual error at the n-th node on the h-th level and,

$$\{f_h^n\} = -\int [B_h^n]'\{\sigma\}dv + \int [A_h^n]'\{g\}dv + \int [A_h^n]'\{\tau\}ds_\sigma \tag{10}$$

$$[K_h^n] = \int [B_h^n]'[D][B_h^n]dv \tag{11}$$

The detail of the solution procedure is as follows.

Step-(1)   As initial values of nodal displacement parameters and stresses, zero is assumed.

Step-(2)   For the n-th node on the 1-st level, solve Eq. (9) to obtain $\{\Delta U_1^n\}$.

Step-(3)   Update $\{U_1^n\}$ using $\{\Delta U_1^n\}$.

$$\{U_1^n\}=\{U_1^n+\Delta U_1^n\} \quad (1<n<N_1) \tag{12}$$

Step-(4)   Using updated $\{U_1\}$, update $\{\sigma\}$ according to Eq. (4).

Step-(5)   From updated $\{\sigma\}$ and Eq. (9), compute $\{\Delta U_2^n\}$.

Step-(6)   Repeat Step-(2) through Step-(5) for levels h=2 to h=H.

Step-(7)   Compute residual error $\{e_h^n\}$ and its norm E_{rror}.

$$\{e_h^n\}=\{f_h^n\}=-\int [B_h^n]^T\{\sigma\}dv+\int [A_h^n]^T\{g\}dv+\int [A_h^n]^T\{\tau\}ds_\sigma \quad (1<n<N_h) \tag{13}$$

$$E_{rror}=\left(\sum_{h=1}^{H}\sum_{n=1}^{N_h}\{e_h^n\}^T\{e_h^n\}\right)^{1/2} \tag{14}$$

Step-(8)   Repeat Step-(2) through Step-(7) until the error norm E_{rror} becomes smaller than the tolerance.

Step-(9)   Substitute the final value of $\{U_h\}$ into Eqs. (2), (3) and (4) to compute displacement u, strain ε and stress σ.

The details of the computational scheme may be different when the problem to be solved is different, such as heterogeneous or nonlinear problems. If the problem is a linear isotropic two or three dimensional problem and the regular uniform mesh is used, the stiffness matrix given by Eq. (9) is identical or similar for all internal nodes regardless of the level of hierarchy. In case of two and three dimensional elastic problems, Eq. (9) becomes simultaneous equations with two and three unknowns, respectively. The variables to be saved during computation, in ideal cases, are nodal displacement parameters $\{\Delta U_h^n\}$ and the error vectors $\{f_h^n\}$. Thus the size of the memory necessary is the order of $2DOF$, where DOF is the total degree of the problem.

TWO DIMENSIONAL EXAMPLE PROBLEM

The following two dimensional elastic problem is taken as an example. The problem is the stretching of a square sheet by the forced displacement at four corners. Figure 2 shows the comparison between the old type FMG and newly developed HMG with respect to the convergence of the error with the iteration. The problem is solved using hierarchical mesh with seven levels (33,274 DOF). Though convergence rate of the new FMG is slower than that of the old one, no fluctuation is observed during convergence and after the convergence as opposed to the old FMG. Figure 3 shows the relation between the DOF of the problem and the computing time to achieve the relative error of 10^{-6}. As it is shown in the figure, a problem with 134,250,490 DOF can be solved by one 64-bit PC in 22,403 seconds.

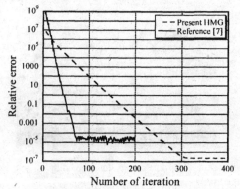

Fig. 2   Convergence with iteration.

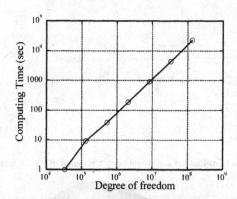

Fig. 3   Relation between *DOF* and computing time.

THREE DIMENSIONAL EXAMPLE PROBLEM

Like standard three dimensional finite element method, the proposed HMG can be regarded as a cubic digital space or a cubic digital clay. The designer or engineer can create any object, such as bioceramics, ceramic part and even ceramic composite structures with complex geometry. The most effective way of using HMG is to generate the model directly from the CAD data or the CT scan digital image such as that of foamed aluminum shown by Fig. 4. The deformation of the foamed aluminum is analyzed by 2-D HMG under the forced displacement along the horizontal axis and the computed result is shown in Fig. 5. Figure 6 shows the deformation of thin steel half spherical shell with circular cutting under thermal load. It is defined in the 256 x 256 x 256 grid space. If the space is fully filled, it consists of 16 million elements. As it is seen from the fact that solution process of HMG is node by node, the computational time for HMG is roughly proportional to the number of elements or nodes. Thus the solution time becomes much less compared with the direct solution method, in which the computational time is proportional to $DOF^{2.3}$ [8], as the size of the problem becomes large.

Fig. 4   Foamed aluminum alloy.

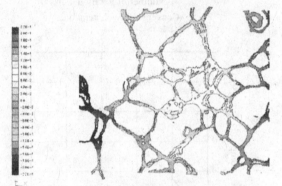

Fig. 5   Displacement of foamed aluminum alloy computed using CT scan data (2D analysis).

Fig. 6   Thermal deformation and stress distribution of a thin half spherical shell
with circular cutting defined in 256 x 256 x 256 grid space
computed by 3D-HMG.

CONCLUSIONS

The iterative solution procedure for the Hierarchical Multi-Grid Method is investigated from the aspect of variational theorem. When the continuity of the displacement field is relaxed, the fractal type Multi-Grid Method may be employed. If the compatibility is assumed, Minimum Potential Energy Theorem can be employed. In this report, the FE code is developed based on the latter and it is shown that the excellent convergence can be maintained without introducing controlling parameters such as the feedback ratio. The robustness of the proposed method indicates its great potential for the application to variety of engineering problems.

REFERENCES

[1]D. Kershaw, "The incomplete Cholesky conjugate gradient method for iterative solution of systems of linear equations," *Journal of Computatinal Physics*, **26**, 43-65 (1978).

[2]P. Geng, J.T. Oden and R.A. van de Geijn, "A Parallel Multifrontal Algorithm and its Imprementation," *Computer Methods in Applied Mechanics and Engineering*, **149**, 289-301 (1997).

[3]W.L. Briggs, "Multi-Grid Methods and Applications," *Springer-Verlag*, Berlin (1985).

[4]K. Stuben, "A Review of Algebraic Multigrid," *Journal of Computational and Applied Mechanics*, **128**, 281-309 (2001).

[5]Q.V. Dinh Glowinski and J. Periaux, "Domain Decomposition Methods for Nonlinear Problems in Fluid Dynamics," *Computer Methods in Applied Mechanics and Engineering*, **40**, 27-109 (1983).

[6]J. Mandel, "Balancing Domain Decomposition," *Communications on Numerical Methods in Engineering*, **9**, 233-341 (1993).

[7]H. Murakawa, M. Tejima, S. Itoh, H. Serizawa, and M. Shibahara, "Ultra Large Scale FE Computation Using Fractal Multi-Grid Method", *Proceedings of the 2005 International Conference on Computational & Experimental Engineering and Sciences (ICCES'05)*, **2**, 228-233 (2005).

[8]H. Murakawa, H. Serizawa, M. Tejima, K. Taguchi and S. Itoh, "Stress Analysis of Geometrically Complex and Ultra Large Scale Model by Fractal Multi-Grid Method", *Ceramic Transactions*, **198**, 21-26 (2007).

OPTICAL METHODS FOR NONDESTRUCTIVE EVALUATION OF SUBSURFACE FLAWS IN SILICON NITRIDE CERAMICS

J. G. Sun, Z. P. Liu
Argonne National Laboratory
Argonne, IL 60439

Z. J. Pei
Kansas State University
Manhattan, KS 66506

N. S. L. Phillips, and J. A. Jensen
Caterpillar Inc.
Peoria, IL 61656

It is known that the strength and lifetime of silicon nitrides are strongly affected by subsurface flaws that are either inherent to the material (voids, porosity, etc.) or induced by component processing such as machining damage (e.g., cracks). Because ceramics are translucent, optical methods are effective to detect and characterize these types of subsurface flaws. In this study, three optical methods were developed/utilized for nondestructive evaluation (NDE) of subsurface flaws in silicon nitride ceramics: (1) laser backscatter, (2) optical coherence tomography (OCT) and, (3) confocal microscopy. The laser backscatter is a two-dimensional method while both OCT and confocal are three-dimensional methods. Subsurface flaws of various types, sizes, and depths can be identified and imaged by these NDE methods. In particular, subsurface Hertzian cracks, induced by surface indentations with various loads, were clearly imaged for the first time by the confocal method. This paper describes these methods and presents NDE data and their correlation with surface photomicrography results.

INTRODUCTION

Advanced ceramics are leading candidates for high-temperature engine applications that offer improved engine performance and reduced emissions. Among these ceramics, silicon nitrides (Si_3N_4) are being evaluated for valve train materials in automotive and diesel engines [1]. However, for high-strength Si_3N_4 ceramics, it is known that surface and subsurface flaws may significantly degrade their fracture strength and fatigue resistance [2-3]. These flaws are normally in the form of microstructural discontinuities that are either inherent material defects (voids, porosity, etc.) or induced by component processing such as machining damage (e.g., spalls, cracks). To ensure the reliability and durability of ceramic components in engine applications, these defects/damages must be detected and characterized.

Because Si_3N_4 ceramics are partially translucent to light and the strength limiting flaws are normally within a shallow depth under the surface, optical methods are effective to detect and characterize these types of subsurface flaws. Argonne National Laboratory (ANL) has developed and utilized several optical methods for nondestructive evaluation (NDE) of subsurface flaws in silicon nitride ceramics. In this study, three optical methods are presented and evaluated: (1) laser backscatter, (2) optical coherence tomography (OCT) and, (3) confocal microscopy. The laser backscatter is a two-dimensional method while both OCT and confocal are three-dimensional methods. It is demonstrated in the following that subsurface flaws of various types, sizes, and depths can be identified and imaged by these NDE methods.

LASER BACKSCATTER

The laser backscatter method utilizes a polarized laser light and cross-polarized backscatter detection to probe the subsurface of an optically translucent ceramic [4-5]. As the laser beam is incident on the air/ceramic interface, a portion of the transmitted light will change polarization states, while the reflected light will not [6]. By detecting only the backscattered light in the cross-polarized direction from the incident beam, subsurface microstructure and flaws can be measured while filtering out all other reflections. Figure 1 illustrates a schematic diagram of the experimental setup of the laser-backscatter scan system. In this system, the cross-polarized backscattered light is measured by two detectors A and B. As the laser illumination is raster scanned over the specimen surface, two gray-scale images are composed: a sum image is calculated from the sum of the measurements of Detectors A and B; and a ratio image is calculated from the ratio of the measurements of Detector B to Detector A. The sum data were found to be most indicative of lateral defects, while the ratio value is more sensitive to median defects [4].

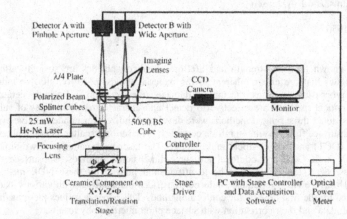

Fig. 1. Illustration of setup for laser scattering inspection system

The laser backscatter system has been used to determine machining induced damage in Si_3N_4 specimens with flat and cylindrical machined surfaces. Figures 2a-b show, respectively, laser-scatter sum and ratio images of the surface of a diamond-ground GS44 Si_3N_4 specimen. The images were obtained with scan pixel size of 10 µm in both directions. In the sum image, the white spots and lines represent surface regions with excessive subsurface backscattering due to surface/subsurface defects or machining damage. Correspondingly, the damaged regions are darker in the ratio image. To identify these defects/damages, the ground surface was polished. Figure 2c is a photomicrograph of the surface after polishing off a 51-µm-thick surface layer. By comparing the laser-scatter images and the photomicrograph, all detected flaw features can be characterized. They include two surface-breaking cracks (identified as "Cracks") at the top-left region of the images, a single prominent spot identified to be a subsurface void, and several line features that were due to machining subsurface damage ("Grounding Damage"). In addition, many relatively weaker spot features distributed in the laser-scatter images represent subsurface regions with higher porosity than surrounding material ("Porous Pores").

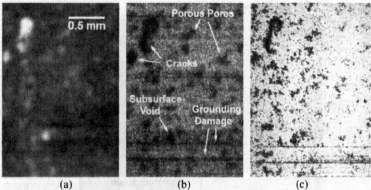

(a)   (b)   (c)

Fig. 2. Laser scatter (a) sum and (b) ratio images of ground surface of a diamond-ground GS44 specimen; and (c) photomicrograph of the surface after polishing.

The laser-scatter system was used to determine the extent of machining damage in transversely ground SN235P Si_3N_4 cylindrical specimens. The specimens was scanned in both axial and longitudinal directions, and the subsurface scattered light intensity at all locations is recorded and processed to compose a two-dimensional scatter image of the scanned region, as shown in Fig. 3. Scan pixel size was 5 µm in both directions. In the laser-scatter (sum) image in Fig. 3b, the gross grayscale variation is likely due to the second-phase inhomogeneity within the Si_3N_4 material and is not considered to be detrimental to the strength [7]. Typical detailed laser-scatter images are shown in Figs. 4a-b for a fine-ground and a rough-ground rod, respectively. The material defects, likely subsurface porous pores appearing as individual high-scattering spots, are seen distributed throughout the surfaces in both rods. The machining damages, represented by brighter horizontal marks ~0.1-mm long along the horizontal grinding direction, are visible only in the image of the rough-ground rod (Fig. 4b). These machining damages are likely median cracks that may cause strength reduction. By combining of laser-scatter NDE data with fractography examination of fracture-tested specimens, it was demonstrated that laser scattering is promising for identifying fracture origins without destructive fracture tests [8].

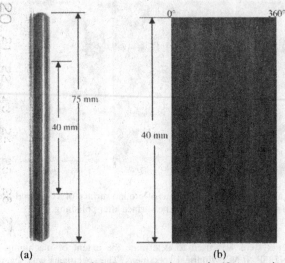

Fig. 3.   (a) Scanned region and (b) laser-scatter image of a 40-mm section of a ceramic rod.

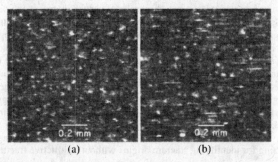

Fig. 4.   Detailed laser-scattering images of (a) fine-ground and (b) rough-ground SN235P rods; the grinding damage is shown as bright horizontal lines in (b).

OPTICAL COHERENCE TOMOGRAPHY (OCT)

OCT is a 3D method originally developed for imaging biological materials. It utilizes a Michelson interferometer to differentiate optical reflection from different depths of a translucent material. The block diagram of the OCT system at ANL is shown in Fig. 5. Light from an optical source is split into two paths, a sample path and a reference path. Light in the reference path is reflected from a fixed-plane mirror, whereas light in the sample path is reflected from surface and subsurface features of a ceramic sample. The reflected light from the sample path will only be detected if it travels a distance that closely matches the distance traveled by the light in the reference path. Thus, by scanning the sample, data can be obtained in a plane perpendicular or parallel to the sample surface. For an OCT system, typical spatial resolution is ~10 μm when a low-coherence diode laser is used.

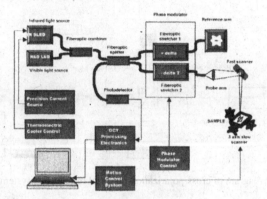

Fig. 5. Block diagram of basic elements in OCT system

Because OCT systems typically use optical sources with longer wavelengths that have deeper optical penetration in ceramics (the ANL system uses a 1310-nm diode laser), they can probe deeper depths into ceramic subsurface than laser-scatter or confocal systems that normally use visible wavelengths. The ANL OCT system was evaluated by scanning the cross section of a NT551 step sample, as shown in Fig. 6. For this sample, the top and bottom surfaces of the two steps of 84- and 184-μm thick are detected (vertical bright stripes are artifacts). The optical thickness of the steps can be directly determined from the images. Because optical penetration depth for OCT equals the physical depth multiplied by the refraction index of the material, direct comparison of optical and physical depths allows for the determination of the refraction index of silicon nitrides. The measured refection index from the Fig. 6 image is 2.0, which is consistent with reported values. In addition, the OCT image reveals many scattering sources within the NT551 subsurface. Higher scattering is usually an indication of porosity or defects; this method may therefore be used for NDE of the Si_3N_4 ceramics.

0.2 mm

Fig. 6. Scanned cross-sectional OCT image of a NT551 step sample.

CROSS-POLARIZATION CONFOCAL MICROSCOPY

Cross-polarization confocal microscopy is a new 3D imaging method developed at ANL [9]. It combines two well-established optical methods, the cross-polarization backscatter detection and the scanning confocal microscopy, and can achieve 3D subsurface imaging with sub-micron spatial resolutions. A schematic diagram of the system is shown in Fig. 7.

Preliminary tests indicated that, with a moderate 40X objective lens at an optical wavelength of 633 nm, the system has an axial (depth) resolution of ~2 μm and a lateral resolution of ~0.6 μm [9].

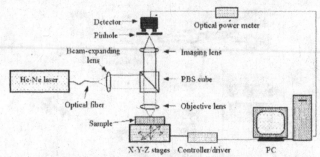

Fig. 7.   Schematic diagram of cross-polarization confocal microscopy system.

A NT551 Si₃N₄ specimen with known subsurface damage was tested using the laser-backscatter and the confocal system. The damage is subsurface Hertzian C cracks induced by indentations with different loading forces. A Hertzian crack has a distinct subsurface profile: it first extends in a cylindrical fashion normal to the surface for some distance and then turns laterally with an angle into the depth [10]. Figure 8a shows a laser-scatter image of a C crack created by a 2400N indentation load. The image clearly shows the crack size, shape, and crack extension direction in the subsurface (scatter intensity decreases with depth).   However, the laser-scatter method cannot resolve the depth profile of the crack under the surface.

The subsurface C crack was examined using the confocal system with a 633-nm-wavelength HeNe laser. Figure 8b shows a lateral (plane) scan image for a portion of the C crack. Higher scatter intensity is observed from the C crack as well as from many individual material defects (porous pores). To examine the crack angle and depth within the subsurface, three cross-sectional scans, spaced at 10-μm distances from each other as marked as Slices #1-#3 in Fig. 8b, were performed. These cross-sectional scan images are shown in Fig. 8c (in each slice the inside of the material is on top and the specimen surface at bottom). The subsurface extension of the C crack (depth and angle) is clearly visualized in these cross-sectional images. It is evident that the subsurface C crack exhibits complex damage patterns that may be the results of its interaction with the material's subsurface microstructure during crack propagation. It is believed that this is the first known result of a direct image of such cracks. The images in Fig. 8c, however, indicate that the detection depth is only ~40 micron for the NT551 material when using a laser wavelength of 633 nm.

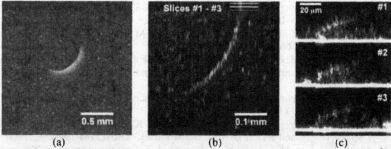

(a) (b) (c)

Fig. 8. (a) Laser-backscatter image of a subsurface C crack; and confocal scan (b) plane image (parallel to surface) and (c) three cross-sectional images (perpendicular to surface) at locations marked in (b).

CONCLUSIONS

Three optical methods were developed/utilized at ANL for nondestructive evaluation of subsurface flaws in silicon nitride ceramics: (1) laser backscatter, (2) optical coherence tomography (OCT) and, (3) confocal microscopy. The principle of and typical data generated from these methods are presented and discussed in this paper. All methods are based on laser scanning of the sample surface and subsurface to construct 2D images that can be used for direct identification of flaws within the sample's subsurface. The laser backscatter is a two-dimensional method without the depth resolution of the subsurface flaws. However, it can quickly build a 2D laser-scatter image of the entire sample surface (i.e., 100% surface inspection), which can be used to identify the type, location, size, and severity of subsurface flaws. Both OCT and confocal methods are three-dimensional methods (i.e., with depth resolution). It was demonstrated that OCT may image the depth >180 μm into the subsurface of a NT551 Si_3N_4 specimen. On the other hand, confocal has a shallower depth penetration, up to ~40 μm deep for NT551 Si_3N_4, but with a better spatial resolution. In particular, the profile of a subsurface Hertzian C crack was clearly imaged for the first time by the confocal method. These optical methods are therefore capable for NDE inspection and characterization of the quality and reliability of ceramic components being considered for high-temperature engine applications.

ACKNOWLEDGMENT

This research was sponsored by the Heavy Vehicle Propulsion Materials Program, DOE Office of FreedomCAR and Vehicle Technology Program, under contract DE-AC05-00OR22725 with UT-Battelle, LLC. Part of the work was supported by the National Science Foundation through the grant DMI-0521203.

REFERENCES

1. J. G. Sun, J. M. Zhang, M. J. Andrews, J. S. Trethewey, N. S. L. Phillips, J. A. Jensen, "Evaluation of Silicon-Nitride Ceramic Valves," Int. J. Appl. Ceram. Tech., 2008, in press
2. R. D. Ott, "Influence of Machining Parameters on the Subsurface Damage of High-Strength Silicon Nitride," Ph.D. Thesis, The University of Alabama at Birmingham, AL, 1997.
3. M. J. Andrews, "Life Prediction and Mechanical Reliability of NT551 Silicon Nitride," Ph.D. Thesis, New Mexico State University, Las Cruces, NM, 1999.

4. J. G. Sun, W. A. Ellingson, J. S. Steckenrider, and S. Ahuja, "Application of Optical Scattering Methods to Detect Damage in Ceramics," in *Machining of Ceramics and Composites*, Part IV, Chapter 19, eds., S. Jahanmir, M. Ramulu, and P. Koshy, Marcel Dekker, New York, pp. 669-699, 1999.
5. W. A. Ellingson, J. A. Todd, and J. G. Sun, "Optical Method and Apparatus for Detection of Defects and Microstructural Changes in Ceramics and Ceramic Coatings," U.S. Patent 6,285,449, issued 2001.
6. J. C. Stover, *Optical Scattering and Analysis*, SPIE Optical Engineering Press, Bellingham, Washington, pp. 111-133 (1995).
7. M. J. Andrews, A. A. Wereszczak, T. P. Kirkland, and K. Breder, "Strength and Fatigue of NT551 Silicon Nitride and NT551 Diesel Exhaust Valves," ORNL/TM-1999/332, 2000.
8. J. M. Zhang, J. G. Sun, M. J. Andrews, A. Ramesh, J. S. Trethewey, and D. M. Longanbach, "Characterization of Subsurface Defects in Ceramic Rods by Laser Scattering and Fractography," in Review of Quantitative Nondestructive Evaluation, eds. D.O. Thompson and D.E. Chimenti, Vol. 25, pp. 1209-1216, 2006.
9. J. G. Sun, "Device and Nondestructive Method to Determine Subsurface Micro-structure in Dense Materials," US Patent No. 7,042,556, issued 2006.
10. B.R. Lawn, *Fracture of Brittle Solids*, Cambridge University Press, New York, 1993.

FRACTOGRAPHIC ANALYSIS OF MINIATURE THETA SPECIMENS

George D. Quinn
Ceramics Division
National Institute of Standards and Technology
Gaithersburg, MD 29899

ABSTRACT

The theta strength test specimen is a simple and elegant way to measure the strength of small structures. Ring shaped test specimens with a central web are compressed diametrally so that the middle web section is stretched in uniform tension. The compression loading scheme eliminates the need for special grips. Prototype miniature silicon specimens with web sections as thin as 7.5 micrometers were fabricated by deep reactive ion etching (DRIE) of single crystal silicon wafers. A conventional nanoindentation machine with a flat head indenter applied load and monitored displacement to fracture. Fractographic analysis of the broken specimens identified fracture mirrors and fracture origins. The origins were 200 nm to 500 nm deep DRIE etch pits.

INTRODUCTION

Mechanical properties at the small scale may play a vital role in design, fabrication, and application of microelectromechanical systems (MEMS) and nanoelectromechanical systems (NEMS) devices. Strength test specimens should be comparable in size as the micro-devices themselves and manufactured in the same manner. Many test specimens are scaled down traditional configurations such as microtensile specimens that are pulled apart by grips or electrostatic devices.[1,2,3,4,5,6,7] Some are cantilever beams that are flexed to fracture. It is challenging to manipulate, install, and apply load to tiny specimens. Many of the testing schemes are prone to the normal misalignments and loading problems associated with testing rigid ceramics and serious experimental errors can occur. A small interlaboratory comparison (round robin) study on miniature polysilicon specimens generated elastic moduli that varied by as much as ± 12 % and average strengths that varied by a factor of two.[6]

One appealing configuration for miniature structures is the theta specimen, invented by A. Durelli and V. Parks in the early 1960s.[8,9] Figure 1 shows a 76 mm diameter Plexiglas specimen made to the size that they recommended. In 2005, we presented preliminary results of our work on miniaturizing this test specimen to 1/250th the original size.[10] We fabricated specimens with both the classic round and a new hexagonal shape as shown in Figure 2. Our subsequent work focused on

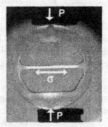

Figure 1. Plexiglas theta specimen. The figure on the right shows the specimen in crossed polarizers while loaded by diametral forces P on the rim. The middle web is stretched in uniform tension σ.

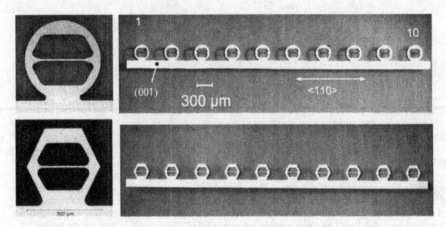

Figure 2. Miniature round and hexagonal theta specimens etched from two silicon wafers. Both are 300 μm wide and have web heights of 7.5 μm and thicknesses of 100 μm.

refining our test procedures to improve the repeatability of our experiments and finite element modeling to analyze the effects of various misalignments and stress concentrations.[10,11] This paper presents new fractographic findings for theta specimens tested to fracture.

MATERIAL

Specimens were fabricated from 100 μm thick double side polished single crystal (100) silicon wafers. Strips of ten specimens were prepared as shown in Figure 2. The round theta specimens were designed to have nominal 10 μm web heights to match Durelli's original shape, but at 1/250th its size. The as fabricated web heights were actually 7.5 μm tall. Hexagonal specimens were also made with 7.5 μm and 22 μm web heights. The wafers were aligned and the strips etched such that the long axis of the strip was parallel to a <110> direction as shown in Figure 2. The specimens were micromachined by deep reactive ion etching (DRIE) with the Bosch process using alternating etch and passivation steps. The sulfur hexafluoride (SF_6) etchant was applied in about 12 s duration steps. The nominal etching rate was 3 μm/min with a standard photoresist type mask. C_4F_8 was used to create the passivation layers between etching steps. About 85 and 100 etch-passivation steps were used to go all the way through the 100 μm wafers for the hexagonal and round specimens, respectively. As will be shown, this conventional treatment created sidewalls with 1 μm bands with 100 nm to 200 nm height undulations. The sidewalls had 1° to 2° re-entrant tapers (undercutting).

EXPERIMENTAL PROCEDURE

Specimens were placed in a holder and loaded to fracture in a nanoindentation machine with a flat-tipped diamond indenter.[a,b] Load and displacement were monitored throughout the loading cycle. The instrument was typically operated in load control with a total ramp time to fracture of about 15 s.

---

[a] Nanoindenter XP, MTS, Oak Ridge, TN.
[b] Certain commercial materials or equipment are identified in this paper to specify adequately the experimental procedure. Such identification does not imply endorsement by the NIST nor does it imply that these materials or equipment are necessarily the best for the purpose.

The load-displacement curves were linear to fracture. Break loads were of the order of 500 mN to 2 N depending upon the specimen type and web size. Between two and eight specimens on a strip were broken at a time. The average strengths ranged from 450 MPa to 600 MPa as described in Ref. 10. Additional information about the specimen holder, other experimental procedures, and the equations to convert break load to fracture stresses are in Ref. 10.

The specimen holder was designed to gently hold a strip in place between parallel glass microscope slides, such that the very tops of the ring specimens were exposed to the nanoindenter head. The glass slides allowed the specimen to be examined both before and after fracture as shown in Figures 3 and 4. Manipulation of the specimen strips and the fractured fragments was done manually with very fine-tip (5 μm radius) tools, precision tweezers, as well as with a mechanical micropositioner with a fine-tip probe. A compound optical microscope was used to photograph and measure the size of selected as-received specimens. All fractures were examined in a stereo binocular microscope at up to 200 power. Selected fragments were carefully harvested and mounted for examination in a field emission scanning electron microscope (FESEM) at up to 200 000 magnification. About fifty ring and hexagonal theta specimens were tested. About forty-five fragments of each type were examined in the FESEM.

The pieces were carefully mounted for the FESEM examination on a small brass nut with one face covered with carbon tape. A common round copper transmission electron microscope grid was placed onto the carbon tape. A single theta fragment placed into each square grid hole in order to keep track of each fragment. Handling and mounting of a remnant theta base strip was easy and was done with fine-tip tweezers. The mounting of a single broken fragment was a much more delicate process. It entailed picking up a fragment (usually held by a small natural electrostatic charge) on the end of a very fine-tip needle and moving the fragment a small distance to the carbon tape. Placement of a fragment onto the carbon tape had to be done very carefully, lest the piece fall in an awkward angle or tip over and lie flat on the tape. The pieces did not need to be coated and were examined with 2 kV or 5 kV accelerating voltage.

Figure 3. Typical fracture outcome. The specimen strip was sandwiched between glass slides. One glass slide was removed after a test to allow the harvesting of the fragments. This image shows the front glass slide partially slid downwards to expose the fragments.

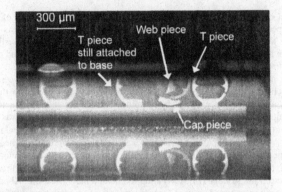

Figure 4. Typical fracture pattern. The two intact and one broken specimens are seen through a front glass slide. The bottom rings are a reflection. Fragments include a rounded cap piece on the bottom, a T piece on the right side, and a triangular web piece.

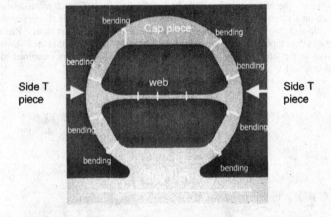

Figure 5. Breakage and fragmentation pattern for the ring theta specimens. The small arrows show the directions of crack propagation

RESULTS

The round theta specimens broke into multiple pieces with several characteristic shapes as shown in Figures 4 and 5. The top round portion often was found as a "cap." The webs broke into several triangular shaped web segments. The sides broke into "T" shaped pieces with either long or short segments at the top of the T. Often one side of a specimen remained attached to the base strip. The pieces that remained attached to the base strip were easy to handle and examine afterwards, but much greater care had to be used with the broken fragments. Only a few triangular web segments were examined since they were very small and difficult to manipulate. The webs almost always broke on {111} type planes creating zigzag or angled fracture surfaces as shown in Figure 6. Sometimes a fracture started on a {110} type plane (Fig. 6d), but it immediately changed to a {111} plane. The {111} is the preferred cleavage plane in single crystal silicon.[12] In one instance, shown in Figure 7, both T sides of a specimen were still on the base after a web section had fractured. The cap piece from the top of the ring broke off and moved to the side.

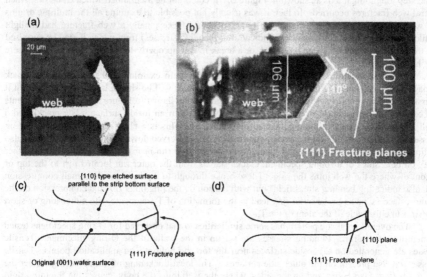

Figure 6. Main fractures in the webs usually ran on {111} type planes. (a) is a front view of a "T" fragment of a hexagonal theta specimen. (b) is a top view looking down onto the web of the same fragment. The fracture planes are marked on the right. (c) and (d) are schematics of the web.

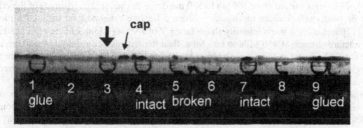

Figure 7. Theta strip after tests to break rings 3 and 6. Rings 2, 5, and 8 had been broken previously. Rings 1 and 9 were tack glued in place to help hold the strip in place. Both sides of ring 3 (large arrow) are still attached to the base, even though the web has fractured out. The top cap of ring 3 broke off and landed to the right (small arrow). Each theta ring has a 300 μm diameter.

FESEM examination of the fracture surfaces revealed the direction of crack propagation and the fracture origins for each fracture plane. The origins were identified by radiating hackle lines and tell tale cathedral fracture mirrors. The direction of crack propagation was identified by general hackle lines, cleavage step hackle, and occasional Wallner lines. Additional information on the fractography of single crystals with explanations and illustrations of these markings may be found in Chapter 8 of the NIST Guide to Fractography of Ceramics and Glasses.[13]

Nearly all web fractures were perpendicular to the web top and bottom surfaces, but at an angle to the web longitudinal axis as shown in Figure 6. It could not be ascertained which came first when several web fractures occurred. In fact, it was usually not possible to examine all the multiple origins and web shard fragments for a particular ring specimen. Occasionally a web fracture had a slight cantilever curl on the fracture plane (an indication of bending stresses) from which it may be surmised that the break was either a secondary break or a break in an improperly-loaded specimen. These were rare, however.

The top cap and side T fragments were much easier to examine, and the origins and crack propagation directions were fairly easy to interpret. Examples will be shown later in this paper, but it is first necessary to characterize the overall breakage pattern shown in Figure 5. The cap fragments had cracks that started and ran perpendicularly outwards from an inside surface. The cracks had a small compression curl as they approached the outer rim. This is a clear manifestation of bending stresses in the ring walls. Similar crack patterns were also observed down at the bottom of a ring, also starting from the inside surface and radiating out to the outer rim near the base strip. On the other hand, on both sides of a round specimen, cracks started from the outer rim located just to the top or bottom of where the web joins the ring. These broke through to the inside with a small compression curl, also indicating bending stress fields but with tension on the ring outer rim and compression on the inside surface. Cracks in the outer ring lead to the formation of T fragments with either long or short segments to either side of the stem of the T.

The overall breakage pattern has some similarities to that observed for O-ring specimens tested diametrally on the rim. Bending stresses are set up in the walls of the O-ring specimen. Tensile stresses are generated on the inside surfaces near the top and bottom load application points. Tensile stresses are also present on the outer side surfaces. The local thickening of our theta rings under the top and bottom load points and on the sides, where the web joins the body, causes the fracture origin locations to shift compared to the O-ring locations.

Figures 8 to 10 show three fracture origins in one T fragment still attached to the base strip. The web fracture is identified by a fracture mirror centered on an etch pit. Figure 11 shows another web fracture. Figure 12 shows other fracture mirrors of various fragments centered on surface etch pits. The origin sizes ranged from 200 nm to 500 nm deep. A rudimentary estimate of the critical flaw size can be made with fracture mechanics. Given a fracture toughness on the {111} plane of 0.8 MPa$\sqrt{m}$ [12], and using a stress intensity shape factor Y of 1.99 for a long surface crack, and a range of average fracture stresses of 450 MPa to 600 MPa, then the critical flaws should be 400 nm to 600 nm deep. It should be borne in mind that this simple estimate ignores that some of the fracture planes are at an angle to the web tensile stress, that some of the secondary fracture origins occurred in areas stressed differently than the web, and some fractures were probably secondary breaks.

The rows of etch pits act as strength limiting flaws. Occasionally the origin was near the back face of the specimen, where etch pitting had transformed part of the specimen into a porous sponge-like structure. The origins were much more commonly about one-quarter to one-third away from that face, near where the etch pit rows just started. Evidently the last etch pits in a row were subjected to greater stress intensification than those in the sponge like region. The DRIE process entails alternating etching-passivating steps that create parallel bands of side wall grooves. The etch pit rows that run perpendicularly to these groove-bands started to form well down into the DRIE troughs. Once the

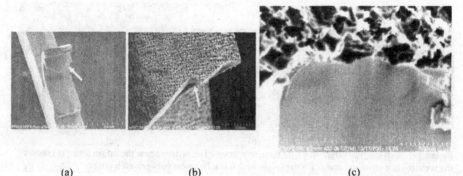

Figure 8. Ring theta specimen, web fracture. (a) shows the entire piece. The arrow shows the origin site. (b) shows a close-up, and (c) shows the fracture mirror and origin flaw, which is an etch pit.

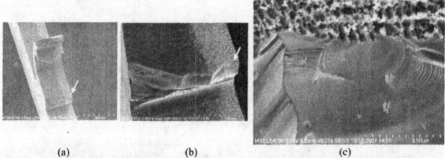

Figure 9. Ring theta specimen, bottom wall fracture. This is the same piece as above, but the origin is an etch pit on the inside surface at the base of the ring wall. A well-defined cathedral fracture mirror is centered on the origin.

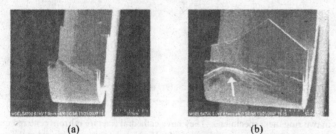

Figure 10. Ring theta specimen, outside wall fracture. This also is the same piece as above, but rotated around. The origin is on the outside wall surface. The fracture plane is very irregular and curved back on the inside surface, indicating a bending stress field.

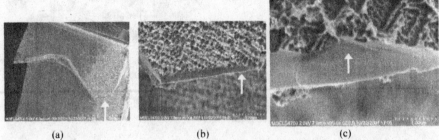

(a)   (b)   (c)

Figure 11. Web fracture origin in a ring theta specimen. The arrows show the origin area. (a) shows the web, (b) is a close-up, and (c) is the origin which is a ligament between etch pits.

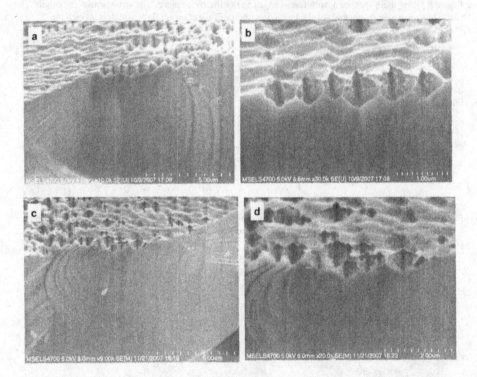

Figure 12. Origin sites in two ring theta specimens. They have cathedral fracture mirrors centered on etch pit flaws. (b) is a close-up of (a), and (d) is a close-up of (c). Note how the origins are at the end of a string of etch pits.

trough becomes fairly deep (about 60 μm through the 100 μm thickness wafer in this instance) the passivation step fails to completely protect a side wall. An etch pit forms and on further processing. the passivation protection layer does not entirely seal it over and subsequent etch steps cause a string or row of new pits to form beneath the first pit.

The hexagonal specimens broke as shown in Figure 13. The typical fragments were web triangles, elbow pieces comprised of the top and one side. and T fragments comprised of a web and two side wall segments. Figures 14 – 16 show web origins. They are all rows of etch pits.

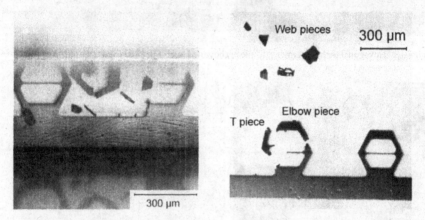

Figure 13. Hexagonal theta specimen fracture illustrating the common fragments. (a) shows the broken pieces seen through the glass slide holder and (b) shows them after extraction.

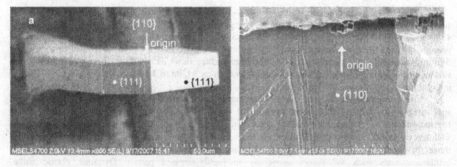

Figure 14. Hexagonal theta specimen web piece showing a web fracture. (a) shows the overall break pattern. The arrows mark the origin on a {110} type plane. (b) is a close-up of (a). Fracture started on a {110} plane but quickly shifted to {111} type planes.

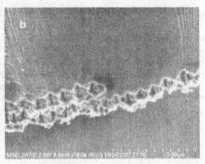

Figure 15. Hexagonal theta specimen "T" piece showing a web fracture. (b) is a close-up of (a).

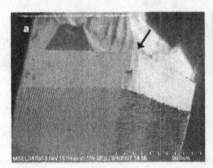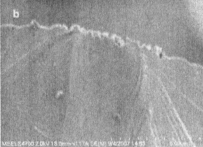

Figure 16. Hexagonal theta specimen "T" piece showing a web fracture. (b) is a close-up of (a).

The 200 nm to 500 nm deep flaws for both the round and hexagonal specimens are among the smallest ever recorded for brittle materials. The average strengths of the specimens were a modest 450 MPa to 600 MPa. Higher strengths for silicon have been reported in many other studies, and presumably the test specimens had even smaller flaws, but good fractographic analysis of tiny strength limiting flaws is rare. Bagdahn et al. [5] observed surface flaws as small as 10 nm to 20 nm on the surface of polysilicon specimens with strengths as high as 2 GPa to 3 GPa. Possible 30 nm – 50 nm internal pore flaws were also identified.

In summary, the fracture origins and breakage patterns in the ring and hexagonal theta specimens were similar. It was not certain that the first breaks were in the web, but the overall breakage patterns are consistent with a scenario in which the webs broke first whereupon secondary fractures occurred from bending stresses in the ring walls. The secondary fracture patterns resemble those for O-ring specimens tested in diametral compression. Secondary fractures are common in brittle materials such as silicon, especially in highly-stressed parts. Dynamic elastic wave reverberations after a web break could trigger the secondary fractures. The dynamic stresses probably do not match a static loading stress solution. Specimen redesign could minimize the incidence of secondary fractures. Further examination will look for evidence of misalignments in these pieces. All fractures initiated at rows of DRIE etch pits that were 200 nm to 500 nm deep. Fracture often commenced at the end of a row. The strengths are on the low end of the range commonly reported for miniature single crystal silicon structures. The surface condition that controlled the strengths in our specimens undoubtedly would control the strengths of actual MEMS components made the same way. Optimized DRIE processing could lead to improved strengths, but that is not the goal of this project, which is to develop a new testing method. We are making new theta and C-ring silicon specimens with better geometry and surface control.

CONCLUSIONS

Micro-sized theta specimens, which are diametrally-compressed ring shaped specimens with a central uniformly stressed web, were used to measure the tension strength of single crystal silicon. The strengths averaged 450 MPa to 600 MPa. The overall fracture patterns were interpreted. The fracture origins were 200 nm to 500 nm deep etch pits that were aligned in rows parallel to the etching direction and perpendicular to the ring faces.

ACKNOWLEDGMENTS

The author thanks Ed Fuller for many helpful discussions about the stress solutions and effects of misalignments. Dan Xiang did many of the strength tests. Jim Beall of NIST-Boulder fabricated the specimens. Grady White helped with many technical discussions and verification of the dimensional accuracy of the specimens.

REFERENCES

[1] S. M. Allameh. "An Introduction to Mechanical-Properties-Related Issues in MEMS Structures," *J. Mat. Sci.*, **38**, 4115-23 (2003).

[2] O. M. Jadaan, N. N. Nemeth, J. Bagdahn, and W. N. Sharpe, Jr., "Probabilistic Weibull Behavior and Mechanical Properties of MEMS Brittle Materials," *J. Mat. Sci.*, **38**, 4087-113 (2003).

[3] T. E. Buchheit, S. J. Glass, J. R. Sullivan, S. S. Mani, D. A. Lavan, T. A. Friedmann, and R. Janek, "Micromechanical Testing of MEMS Materials," *J. Mat. Sci.*, **38**, 4081-86 (2003).

[4] W. N. Sharpe, Jr., D. A. LaVan, and R. L. Edwards, "Mechanical Properties of LIGA-Deposited Nickel for MEMS Transducers," in *Proc. Transducers '97*, Chicago, IL, 607-10 (1997).

[5] J. Bagdahn, W. N. Sharpe, Jr., and O. Jadaan, "Fracture Strength of Polysilicon at Stress Concentrations," *J. Microelectromechanical Systems*, **12**, 302-12 (2003).

[6] W. N. Sharpe, Jr., S. Brown, G. C. Johnson, and W. Knauss, "Round-Robin Test of Modulus and Strength of Polysilicon," *Proc. Mat. Res. Soc.*, **518**, 57-65 (1998).

[7] A. M. Fitzgerald, R. S. Iyer, R. H. Dauskardt, and T. W. Kenny, "Subcritical Crack Growth in Single Crystal Silicon Using Micromachined Specimens," *J. Mater. Res.*, **17** [3] 683-91 (2002).

[8] A. J. Durelli, S. Morse, and V. Parks, "The Theta Specimen for Determining Tensile Strength of Brittle Materials," *Mat. Res. and Stand.*, **2**, 114-7 (1962).

[9] A. J. Durelli and V. J. Parks, "Influence of Size and Shape on the Tensile Strength of Brittle Materials," *Brit. J. Appl. Phys*, **18**, 387-8 (1967).

[10] G. D. Quinn, E. R. Fuller, D. Xiang, A. Jillavenkatessa, L. Ma, D. Smith, and J. Beall, "A Novel Test Method for Small Scale Structures: The Theta Specimen," *Ceram. Eng. and Sci. Proc.*, **26** [2] 117 – 126 (2005).

[11] E. R. Fuller, Jr., D. L. Henann, and L. Ma, "Theta-like Specimens for Measuring Mechanical Properties at the Small-Scale: Effects of Non-Ideal Loading," *Int. J. Mat. Res.*, **98** [8] 729–34 (2007).

[12] C. P. Chen and M. H. Leipold, "Fracture Toughness of Si," *Am. Ceram. Soc. Bull.*, **59** [4] 469-72 (1980).

[13] G. D. Quinn, Guide to Practice for Fractography of Ceramics and Glasses, NIST Special Publication SP 960-16, National Institute of Standards and Technology, Gaithersburg, MD, May 2007.

Author Index